AF553530

TEXTBOOK OF CHEMICAL KINETICS

TEXTBOOK OF CHEMICAL KINETICS

By

Dr. E.H.El-Mossalamy

Chairman/Professor
Department of Chemistry
Faculty of Science
King Abdul Aziz University
Jeddah, (K.S.A.)

&

Dr. Syed Aftab Iqbal

M.Sc. , Ph.D., FICS
FICC, FIAEM, MNASc.
Professor
Department of Chemistry
Saifia Science College
Barkatullah University
Bhopal (India)

DISCOVERY PUBLISHING HOUSE PVT. LTD.

NEW DELHI-110 002

Published by:
Tilak Wasan

DISCOVERY PUBLISHING HOUSE PVT. LTD.
4831/24, Ansari Road, Prahlad Street
Darya Ganj, New Delhi-110002 (India)
Phone: +91-11-23279245, 43764432
Fax: +91-11-23253475
E-mail: parul.wasan@gmail.com
discoverypublishinghouse@gmail.com
info@discoverypublishinggroup.com
web: www.discoverypublishinggroup.com

***First Edition:* 2011**
ISBN: 978-81-8356-833-3

Textbook of Chemical Kinetics

Printed at:
Shree Balaji Art Press
Delhi

Preface

Chemical kinetics, also known as reaction kinetics, is the study of rates of chemical processes. Chemical kinetics includes investigations of how different experimental conditions can influence the speed of a chemical reaction and yield information about the reaction's mechanism and transition states, as well as the construction of mathematical models that can describe the characteristics of a chemical reaction. In 1864, Peter Waage and Cato Guldberg pioneered the development of chemical kinetics by formulating the law of mass action, which states that the speed of a chemical reaction is proportional to the quantity of the reacting substances.

Chemical kinetics deals with the experimental determination of reaction rates from which rate laws and rate constants are derived. Relatively simple rate laws exist for zero order reactions (for which reaction rates are independent of concentration), first order reactions, and second order reactions, and can be derived for others. In consecutive reactions the rate-determining step often determines the kinetics. In consecutive first order reactions, a steady state approximation can simplify the rate law. The activation energy for a reaction is experimentally determined through the Arrhenius equation and the Eyring equation. The main factors that influence the reaction rate include: the physical state of the reactants, the concentrations of the reactants, the temperature at which the reaction occurs, and whether or not any catalysts are present in the reaction.

—Author

Contents

1

Introduction

Chemical kinetics, also known as 'reaction kinetics', is the study of rates of chemical processes. Chemical kinetics includes investigations of how different experimental conditions can influence the speed of a chemical reaction and yield information about the reaction's mechanism and transition states, as well as the construction of mathematical models that can describe the characteristics of a chemical reaction. In 1864, Peter Waage and Cato Guldberg pioneered the development of chemical kinetics by formulating the law of mass action, which states that the speed of a chemical reaction is proportional to the quantity of the reacting substances.

Rate of Reaction

Chemical kinetics deals with the experimental determination of reaction rates from which rate laws and rate constants are derived. Relatively simple rate laws exist for zero order reactions (for which reaction rates are independent of concentration), first order reactions, and second order reactions, and can be derived for others. In consecutive reactions the rate-determining step often determines the kinetics. In consecutive first order reactions, a steady state approximation can simplify the rate law. The activation energy for a reaction is experimentally determined through the Arrhenius equation and the Eyring equation. The main factors that influence the reaction rate include: the physical state of the reactants, the concentrations of the reactants, the temperature at which the reaction occurs, and whether or not any catalysts are present in the reaction.

Generic potential energy diagram showing the effect of a catalyst in an hypothetical exothermic chemical reaction. The presence of the catalyst opens a different reaction pathway with a lower activation energy. The final result and the overall thermodynamics are the same.

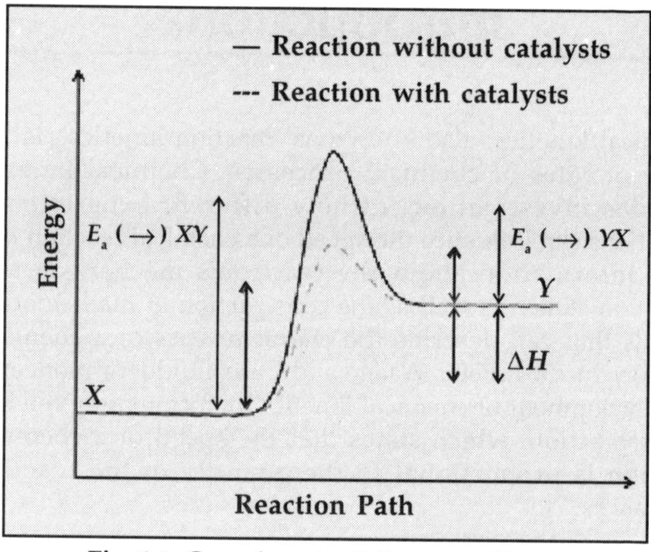

Fig. 1.1: Generic potential energy diagram

Nature of the Reactants

Depending upon what substances are reacting, the time varies. Acid reactions, the formation of salts, and ion exchange are fast reactions. When covalent bond formation takes place between the molecules and when large molecules are formed, the reactions tend to be very slow.

Physical State

The physical state (solid, liquid, or gas) of a reactant is also an important factor of the rate of change. When reactants are in the same phase, as in aqueous solution, thermal motion brings them into contact. However, when they are in different phases, the reaction is limited to the interface between the reactants. Reaction can only occur at their area of contact, in the case of a liquid and a gas, at the surface of the liquid. Vigorous shaking and stirring may be needed to bring the reaction to completion. This means that the more finely divided a solid or liquid reactant, the greater its surface area per unit volume, and the more contact it makes with the other reactant, thus the faster the reaction. To make an analogy, for example, when one starts a fire, one uses wood chips and small branches—one doesn't start with large logs right away. In organic chemistry On water reactions are the exception to the rule that homogeneous reactions take place faster than heterogeneous reactions.

Concentration

Concentration plays a very important role in reactions according to the collision theory of chemical reactions, because molecules must collide in order to react together. As the concentration of the reactants increases, the frequency of the molecules colliding increases, striking each other more frequently by being in closer contact at any given point in time. Think of two reactants being in a closed container. All the molecules contained within are colliding constantly. By increasing the amount of one or more of the reactants it causes these collisions to happen more often, increasing the reaction rate.

Temperature

Temperature usually has a major effect on the rate of a chemical reaction. Molecules at a higher temperature have more thermal energy. Although collision frequency is greater at higher temperatures, this alone contributes only a very small proportion to the increase in rate of reaction. Much more important is the fact that the proportion of reactant molecules with sufficient energy to react (energy greater than activation energy: $E > E_a$) is significantly higher and is explained in detail by the Maxwell–Boltzmann distribution of molecular energies.

The 'rule of thumb' that the rate of chemical reactions doubles for every 10° C temperature rise is a common misconception. This may have been generalized from the special case of biological systems, where the Q_{10} (temperature coefficient) is often between 1.5 and 2.5.

A reaction's kinetics can also be studied with a temperature jump approach. This involves using a sharp rise in temperature and observing the relaxation rate of an equilibrium process.

Catalysts

A catalyst is a substance that accelerates the rate of a chemical reaction but remains chemically unchanged afterwards. The catalyst increases rate reaction by providing a different reaction mechanism to occur with a lower activation energy. In autocatalysis a reaction product is itself a catalyst for that reaction leading to positive feedback. Proteins that act as catalysts in biochemical reactions are called enzymes. Michaelis-Menten kinetics describe the rate of enzyme mediated reactions.

In certain organic molecules specific substituents can have an influence on reaction rate in neighbouring group participation.

Agitating or mixing a solution will also accelerate the rate of a chemical reaction, as this gives the particles greater kinetic energy, increasing the number of collisions between reactants and therefore the possibility of successful collisions.

Increasing the pressure in a gaseous reaction will increase the number of collisions between reactants, increasing the rate of reaction. This is because the activity of a gas is directly proportional to the partial pressure of the gas. This is similar to the effect of increasing the concentration of a solution. A catalyst does not affect the position of the equilibria, as the catalyst speeds up the backward and forward reactions equally.

Equilibrium

While chemical kinetics is concerned with the rate of a chemical reaction, thermodynamics determines the extent to which reactions occur. In a reversible reaction, chemical equilibrium is reached when the rates of the forward and reverse reactions are equal and the concentrations of the reactants and products no longer change. This is demonstrated by, for example, the Haber–Bosch process for combining nitrogen and hydrogen to produce ammonia. Chemical clock reactions such as the Belousov–Zhabotinsky reaction demonstrate that component concentrations can oscillate for a long time before finally attaining the equilibrium.

Free Energy

In general terms, the free energy change (ΔG) of a reaction determines if a chemical change will take place, but kinetics describes how fast the reaction is. A reaction can be very exothermic and have a very positive entropy change but will not happen in practice if the reaction is too slow. If a reactant can produce two different products, the thermodynamically most stable one will generally form except in special circumstances when the reaction is said to be under kinetic reaction control. The Curtin–Hammett principle applies when determining the product ratio for two reactants interconverting rapidly, each going to a different product. It is possible to make predictions about reaction rate constants for a reaction from free-energy relationships.

The kinetic isotope effect is the difference in the rate of a chemical reaction when an atom in one of the reactants is replaced by one of its isotopes.

Chemical kinetics provides information on residence time and heat transfer in a chemical reactor in chemical engineering and the molar mass distribution in polymer chemistry.

Applications

The mathematical models that describe chemical reaction kinetics provide chemists and chemical engineers with tools to better understand and describe chemical processes such as food decomposition, microorganism growth, stratospheric ozone decomposition, and the complex chemistry of biological systems. These models can also be used in the design or modification of chemical reactors to optimize product yield, more efficiently separate products, and eliminate environmentally harmful by-products. When performing catalytic cracking of heavy hydrocarbons into gasoline and light gas, for example, kinetic models can be used to find the temperature and pressure at which the highest yield of heavy hydrocarbons into gasoline will occur.

2

Reaction Rates

Chemical kinetics is the branch of chemistry which addresses the question: "how fast do reactions go?" Chemistry can be thought of, at the simplest level, as the science that concerns itself with making new substances from other substances. Or, one could say, chemistry is taking molecules apart and putting the atoms and fragments back together to form new molecules. All of this is to say that chemical reactions are the core of chemistry.

If Chemistry is making new substances out of old substances (i.e., chemical reactions), then there are two basic questions that must be answered:

1. Does the reaction want to go? This is the subject of chemical thermodynamics.
2. If the reaction wants to go, how fast will it go? This is the subject of chemical kinetics.

Here are some examples. Consider the reaction,

$2\,H_2(g) + O_2(g) = 2H_2O(l)$

We can calculate $_rG^{\circ}$ for this reaction from tables of free energies of formation (actually this one is just twice the free energy of formation of liquid water). We find that $_rG^{\circ}$ for this reaction is very large and negative, which means that the reaction wants to go very very strongly. A more scientific way to say this would be to say that the equilibrium constant for this reaction is very very large.

However, we can mix hydrogen gas and oxygen gas together in a blub or other container, even in their correct stoichiometric proportions, and they will stay there for centuries, perhaps even forever, without reacting. (If we drop in a catalysts - say a tiny piece of platinum - or introduce a spark, or even illuminate the mixture with sufficiently high frequency uv light, or compress and heat the mixture, the mixture will explode.) The problem is not that the reactants do not want to form the products, they do, but they cannot find a "pathway" to get from reactants to products.

Another example: consider the reaction

C (*diamond*) C(*graphite*)

If you calculate $_rG^{\circ}$ or this reaction from data in the tables of thermodynamic properties you will find once again that it is negative (not very large, but still negative). This result tells us that diamonds are thermodynamically unstable. Yet diamonds are highly regarded as gem stones ("diamonds are forever") and are considered by some financial advisors as a good long-term investment hedge against inflation. On the other hand, if you were to vaporize a diamond in a furnace, under an inert atmosphere, and then condense the vapor, the carbon would come back as graphite and not as diamond.

How can all these things be?

The answer is that thermodynamics is not the whole story in chemistry. Not only do we have to know whether a reaction is thermodynamically favored, we also have to know whether the reaction can or will proceed at a finite rate. The study of the rate of reactions is called chemical kinetics.

The study of chemical kinetics requires new definitions, new types of experimental data, and new theories and equations to organize the data. We begin with the definition of *reaction rate.*

Reaction Rates

Consider the reaction,

$2\ NO(g) + O_2(g)\ \ 2\ NO_2(g)$

We can specify the rate of this reaction by telling the rate of change of the partial pressures of one the gases. However, it is convenient to convert these pressures into concentrations, so we will write our rates and rate equations in terms of concentrations, where square brackets, [ ], mean concentration in mol/L.

We might try to write the rate variously as,

$$\frac{d[NO_2]}{dt},$$

or as

$$\frac{d[O_2]}{dt},$$

but these are not the same because each molecule of O_2 gives two molecules of NO_2. To arrive at an unambiguous definition of reaction rate we define the "reaction velocity," v, as

$$v = \frac{1}{2}\frac{d[NO_2]}{dt} = -\frac{d[O_2]}{dt} = -\frac{1}{2}\frac{d[NO]}{dt} \tag{2.1}$$

This is unambiguous. The negative sign tells us that that species is being consumed and the fractions take care of the stoichiometry. Any one of the three derivatives can be used to define the rate of the reaction.

For a general reaction,

$$aA + bB = cC + dD, \tag{2.2}$$

the reaction velocity can be written in a number of different but equivalent ways,

$$v = \frac{1}{a}\frac{d[A]}{dt} = -\frac{1}{b}\frac{d[B]}{dt} = -\frac{1}{c}\frac{d[C]}{dt} = \frac{1}{d}\frac{d[D]}{dt} \tag{2.3}$$

As in our previous example, the negative signs account for material that is being consumed in the reaction and the positive signs account for material that is being formed in the reaction. The stoichiometry is preserved by dividing the rate of change of concentration of each substance by its stoichiometric coefficient.

Rate Laws

A rate law is an equation that tells us how fast the reaction proceeds and how the reaction rate depends on the concentrations of the chemical species involved. A rate law is an equation of the form,

$$v = f\left([A],[B],\ldots[E]\right). \tag{2.4}$$

Equation (2.4) is gives us a first order differential equation in t because the reaction velocity is related to a time-derivative of one of the concentrations (as in Eq. 2.3).

The rate law may contain substances which are not in the balanced reaction and may not contain some things that are in the balanced equation (even on the reactant side).

Usually rate laws take the form,

$$v = k\left([A]^x, [B]^y [E]^z\right)...., \tag{2.5}$$

where x, y, z, are small whole numbers or simple fractions and k is called the "rate constant." The sum of $x + y + z +$ . . . is called the "order" of the reaction.

Common Types of Rate Laws

1. *First Order Reactions*

In a first order reaction the rate is proportional to the concentration of one of the reactants. That is,

$$v = \text{rate} = k[\text{B}], \tag{2.6}$$

where B is a reactant. If we have a reaction which is known to be first order in B, such as B + other reactants products, we should write the rate law as,

$$-\frac{d[B]}{dt} = k[B] \tag{2.7}$$

The constant, k, in this rate equation is the first order rate constant.

2. *Second Order Reactions*

In a second order reaction the rate is proportional to concentration squared. For example, possible second order rate laws might be written as

$$\text{Rate} = k[\text{B}]^2 \tag{2.8}$$

or as

$$\text{Rate} = \text{k}[\text{A}]\ [\text{B}] \tag{2.9}$$

That is, the rate might be proportional to the square of the concentration of one of the reactants, or it might be proportional to the product of two different concentrations.

3. *Third Order Reactions*

There are several different ways to write a rate law for a third order reaction. One might have cases where

$$\text{Rate} = k[\text{A}]^3, \tag{2.10}$$

or

$$\text{Rate} = k[\text{A}]^2\ [\text{B}] \tag{2.11}$$

or

Rate = k[A] [B] [C] (2.12)

and so on.

We will see later that there are other, more "interesting" rate laws in nature, but a large fraction of rate laws will fit in one of the above categories.

Integrated Forms of Rate Laws

In order to understand how the concentrations of the species in a chemical reaction change with time it is necessary to integrate the rate law (which is given as the time-derivative of one of the concentrations) to find out how the concentrations change over time.

1. *First Order Reactions*

Suppose we have a first order reaction of the form,

B + ... products. (2.13)

Then we can write the rate law and integrate it as follows (recall that the derivative is negative because the concentration of the reactant, B, is decreasing):

$$\frac{d[B]}{dt} = -k[B] \tag{2.14a}$$

$$\frac{d[B]}{[B]} = -k[dt] \tag{2.14b}$$

$$\ln\frac{[B]}{[B]_0} = -kt \tag{2.14c}$$

$$[B] - [B]_0\, e^{-kt} \tag{2.14d}$$

The first order rate law is a very important rate law, radioactive decay and many chemical reactions follow this rate law and some of the language of kinetics comes from this law. The form of Eq. (2.14d) is called an "exponential decay." This form appears in many places in nature. One of its consequences is that it gives rise to a concept called "half-life."

Half-life

The half-life, usually symbolized by $t_{½}$, is the time required for [B] to drop from its initial value $[B]_o$ to $[B]_o/2$.

Using the integrated form of the first order rate law we find that

$$\frac{[B]_0}{2} = [B]_0 e^{-kt½} \tag{2.15}$$

or

$$½ = e^{-kt½}$$

Taking the logarithm of both sides gives,

$$\ln\frac{1}{2} = -kt_{½} \tag{2.16a}$$

$$\ln 2 = kt_{½} \tag{2.16b}$$

or

$$t_{½} = \frac{\ln 2}{k} \tag{2.17}$$

You can also write

$$[B] = [B]_o 2^{-\frac{t}{t_{1/2}}} = \frac{[B]_o}{2^{\frac{t}{t½}}} \tag{2.18}$$

which may actually give a little more insight into what is meant by half-life. This equation demonstrates clearly that the concentration drops by a factor of two for every $t_{1/2}$ increment in time.)

For first order processes it is common to define a "relaxation time." t , by

$$t = \frac{1}{k} \tag{2.19}$$

so that one can write the integrated form of the rate law as

$$[\mathrm{B}] = [\mathrm{B}]_0 e^{-t/r} \tag{2.20}$$

t is the time required for [B] to drop from $[\mathrm{B}]_o$ to $[\mathrm{B}]_o/e$. Sometimes t is called the "one over e" time. Although the half-life is almost always used to describe the decay rate of radioactive elements, it is common for chemists to talk about the rate of first order processes in chemistry in terms of the relaxation time.

2. *Simple Second Order Rate Equations*

A typical simple second order reaction, where B is one of the reactants, would look like,

$$\frac{d[\mathrm{B}]}{dt} = -k[\mathrm{B}]^2 \tag{2.21}$$

which can be rearranged for integration as,

$$\frac{d[\mathrm{B}]}{[\mathrm{B}]^2} = -kdt \tag{2.22}$$

This equation can be integrated to give,

$$-\frac{1}{[\mathrm{B}]} + \frac{1}{[\mathrm{B}]_0} = -kt \tag{2.23}$$

or

$$[B] = \frac{1}{\frac{1}{[B]_0} + kt} \tag{2.24}$$

We can make a quick check to see if this result fits what we know about the system. At $t = 0$ we get $[B] = [B]_o$ which is correct and at t = infinity we find that $[B] = 0$ which is also what we would expect.

Half-life

We can define the half-life of a second order reaction in the same way as in first order reactions. That is, the half-life, $t_{1/2}$, is the time required for [B] to fall from $[B]_o$ to $[B]_o/2$. Thus we use Eq. (2.23) to give

$$-\frac{1}{\frac{[B]_0}{2}} + \frac{1}{[B]_0} = -kt_{1/2} \tag{2.25a}$$

or

$$-\frac{1}{[B]} = -kt_{1/2} \tag{2.25b}$$

and finally

$$t_{1/2} = \frac{1}{k[B]_0} \tag{2.26}$$

Note that the half-life for a simple second order reactions depends on the initial concentration and that it is proportional to $1/[B]_o$. Contrast this to what we know about first order reactions where the half-life is independent of the initial concentration. This fact may come in useful later when we try to find experimental methods for determining rate laws.

3. *Mixed Second Order Rate Equations*

A mixed second order rate law equation, where both components A and B are reactants, would look like,

$$\frac{d[\mathrm{B}]}{dt} = -k[\mathrm{B}][\mathrm{A}] \tag{2.27}$$

Eq. (2.27) has two variables in it (besides time). In order to integrate this equation we need to know the stoichiometry so that we can tell how [A] depends on [B].

We will use the following stoichiometry as an example,

$$2\mathrm{A} + \mathrm{B} \quad \mathrm{P} \; + \text{etc} \quad (\mathrm{P} = \text{product}) \tag{2.28}$$

We have a choice as to which component concentration to use for our variable, we can use [A], [B], or [P].

Let's try [P] as our variable. (The form of the answer will depend on the choice of variable. A different choice of variable will give an answer that looks different even though it is algebraically equivalent, see below.)

Let's assume that there is no product present at the beginning of the experiment, that is,

$$[\mathrm{P}]_o = 0 \tag{2.29}$$

We now need to express [A] and [B] in terms of [P]. We can see that,

$$[\mathrm{A}] = [\mathrm{A}]_o \; - 2\,[\mathrm{P}] \tag{2.30}$$

(We obtain Eq. (2.30) from the stoichiometry, which says that every time we make one mole of P we use up two moles of A. That is, each P formed takes away two A's.) Also,

$$[\mathrm{B}] = [\mathrm{B}]_o \; - [\mathrm{P}] \tag{2.31}$$

because each P formed takes away only one B.

So our rate equation in terms of the single variable, [P] becomes,

$$\frac{d[\mathrm{B}]}{dt} = -\frac{d[\mathrm{P}]}{dt} \tag{2.32}$$

and then

$$\frac{d[\mathrm{P}]}{dt} = -k\left([\mathrm{A}]_0 - 2[\mathrm{P}]\right)\left([\mathrm{B}]_0 - [\mathrm{P}]\right) \tag{2.33}$$

In order to make this eq. (2.33) easier to integrate we rearrange it as follows:

$$-\frac{d[\mathrm{P}]}{dt} = -2k\left([\mathrm{P}] - \frac{[\mathrm{A}]_0}{2}\right)\left([\mathrm{P}] - [\mathrm{B}]_0\right) \tag{2.34}$$

or

$$-\frac{d[\mathrm{P}]}{\left([\mathrm{P}] - \frac{[\mathrm{A}]_0}{2}\right)\left([\mathrm{P}] - [\mathrm{B}]_0\right)} = -2k\,dt \tag{2.35}$$

The left-hand side can be integrated by partial fractions, or you can look it up in a book of integral tables. Since this is a common problem in chemical kinetics we will give a brief review of the method of partial fractions.

Review of Partial Fractions

We would like to write the fraction appearing on the left side of Equation 15 as a sum of two simpler fractions. We can do this if we can find two quantities, X and Y, such that,

$$\frac{1}{\left([\mathrm{P}] - \frac{[\mathrm{A}]_0}{2}\right)\left([\mathrm{P}] - [\mathrm{B}]_0\right)} = \frac{X}{\left([\mathrm{P}] - \frac{[A]_0}{2}\right)} + \frac{Y}{\left([\mathrm{P}] - [\mathrm{B}]_0\right)} \tag{2.36}$$

Put the right-hand side over the common denominator.,

$$\frac{1}{\left([P]-\frac{[A]_0}{2}\right)\left([P]-[B]_0\right)} = \frac{X[P]-[B]_0 + Y\left([P]-\frac{[A]_0}{2}\right)}{\left([P]-\frac{[A]_0}{2}\right)\left([P]-[B]_0\right)} \quad (2.37)$$

Notice that both sides have the same denominator. Then the numerators must be equal,

$$1 = X\left([P]-[B]_0\right) + Y\left([P]-\frac{[A]_0}{2}\right) \quad (2.38)$$

or

$$1 = X[P] + Y[P] - X[B]_0 - Y\frac{[A]_0}{2} \quad (2.39)$$

Since in Eq. (2.39) the left-hand side, 1, does not depend on [P] we conclude that the [P] on the right-hand side must cancel out. That is,

$$X[P] + Y[P] = 0 \quad (2.40)$$

Therefore

$$Y = -X \quad (2.41)$$

We also include that

$$-X[B]_0 - Y\frac{[A]_0}{2} = 1 \quad (2.42)$$

or, using Eq. (2.41)

$$-X[B]_0 + X\frac{[A]_0}{2} = 1 \quad (2.43)$$

so

$$X = \frac{1}{\frac{[A]_0}{2} - [B]_0} \qquad Y = \frac{-1}{\frac{[A]_0}{2} - [B]_0} \tag{2.44a, b}$$

Then, inserting this X and Y back into Equation 13, we get,

$$\frac{1}{\left([P] - \frac{[A]_0}{2}\right)\left([P] - [B]_0\right)} = \frac{1}{\left([P] - \frac{[A]_0}{2}\right)\left(\frac{[A]_0}{2} - [B]_0\right)} - \frac{1}{\left([P] - [B]_0\right)\left(\frac{[A]_0}{2} - [B]_0\right)} \tag{2.45}$$

——————————— (End of review of partial fractions.)

which we can substitute back into Eq. (2.35) to get

$$\frac{d[P]}{\left([P] - \frac{[A]_0}{2}\right)\left(\frac{[A]_0}{2} - [B]_0\right)} - \frac{d[P]}{\left([P] - [B]_0\right)\left(\frac{[A]_0}{2} - [B]_0\right)} = 2kdt \tag{2.46}$$

Although this is messy, it is relatively easy to integrate to give

$$\frac{1}{\left(\frac{[A]_0}{2} - [B]_0\right)} \ln \frac{\left([P] - \frac{[A]_0}{2}\right)}{-\frac{[A]_0}{2}} - \frac{1}{\left(\frac{[A]_0}{2} - [B]_0\right)} \ln \frac{([P] - [B]_0)}{-[B]_0} = 2kt \tag{2.47}$$

Change sign inside the logarithms and combine them

$$\frac{1}{\left(\frac{[A]_0}{2} - [B]_0\right)} \ln \frac{\left(\frac{[A]_0}{2} - [P]\right)}{-\frac{[A]_0}{2}} - \frac{1}{\left(\frac{[A]_0}{2} - [B]_0\right)} \ln \frac{([B]_0 - [P])}{-[B]_0} = 2kt \tag{2.48}$$

$$\frac{1}{\left(\frac{[A]_0}{2} - [B]_0\right)} \ln \frac{\left(\frac{[A]_0}{2} - [P]\right)[B]_0}{\left([B]_0 - [P]\right)\frac{[A]_0}{2}} = 2kt \tag{2.49}$$

and then,

$$\ln \frac{\left(\frac{[A]_0}{2} - [P]\right)[B]_0}{\left([B]_0 - [P]\frac{[A]_0}{2}\right)} = \left(\frac{[A]_0}{2} - [B]_0\right)2kt \tag{2.50}$$

This result is probably good enough for our purposes, but you can, in fact, solve for [P],

$$[P] = \frac{\frac{[A]_0}{2}[B]_0\left(e[A]_0 kt - e^2[B]_0 kt\right)}{\frac{[A]_0}{2}e[A]_0 kt - [B]_0 e^2[B]_0 kt} \tag{2.51}$$

Note that $[A]_o = 2\ [B]_o$ is a special case. (Stoichiometrically equivalent amounts.) If you try to plug these initial concentrations into Eq. (2.51) for [P] you get 0/0. Mathematically you can make this work by taking the limit as $[A]_o\ 2\ [B]_o$, but this is messy. It is easier to go back to the beginning and convert.

$$-\frac{d[P]}{dt} = -2k\left([P] - \frac{[A]_0}{2}\right)\left([P] - [B]_0\right) \tag{2.52}$$

into

$$-\frac{d[P]}{dt} = -2k\left([P] - [B]_0\right)\left([P] - [B]_0\right) = -2k\left([P] - [B]_0\right)^2 \tag{2.53}$$

This is just a simple second order rate equation which we can set up to integrate as,

$$-\frac{d[P]}{\left([P] - [B]_0\right)^2} = -2kdt \tag{2.54}$$

which integrates to,

$$-\frac{1}{([P]-[B]_0)}+\frac{1}{-[B]_0}=2kt \tag{2.55}$$

or

$$\frac{1}{([B]_0-[P])}-\frac{1}{[B]_0}=2kt \tag{2.56}$$

Eq. (2.56) can easily be solved for [P] and we can determine an expression for the half-life as we have done before

Alternative Solution

We indicated above that we have a choice of variables in the integration of the mixed second order rate Eq. (2.7). Different choices give results that look different on the surface, but which must be algebraically equivalent. Let's integrate Eq. (2.7) again with the same stoichiometry, but this time using [B] as the variable. Our stoichiometry is determined by the balanced reaction,

2A + B P + etc (P = product)

from which we conclude that the amount of B which is used up is given by,

$$\text{B used} = [B]_o - [B] \tag{2.57}$$

Since each B used takes 2A along with it, the amount of A used is

$$\text{A used} = 2 \times (\text{amount of B used}) = 2\,([B]_o - [B]) \tag{2.58}$$

Therefore, the concentration of A is,

$$[A] = [A]_o - \text{A used} = [A]_o - 2[B]_o + 2[B] \tag{2.59}$$

(Notice that [A] in Eq. (2.59) can never be negative because a negative concentration is not physically possible.

If the reactants are not present in their stoichiometrically equivalent amounts we would have to worry about which reagent was the "limiting reagent," that is, which one we will run out of first. For example, if $2[B]_o$ is larger than $[A]_o$ the reaction would cease at the point when Eq. (2.59) reached zero.)

Our mixed second order rate law was,

$$\frac{d[B]}{dt} = -k[B][A]$$

Using Eq. (2.59) for [A] we get

$$\frac{d[B]}{dt} = -k[B][A]$$

$$= -k[B]([A]_o - 2[B]_o + 2[B]) \tag{2.60}$$

$$= 2k[B]\left([B] + \frac{[A]_o}{2} - [B]_o\right)$$

We set this up to integrate as,

$$\frac{d[B]}{[B]\left([B] + \frac{[A]_0}{2} - [B]_0\right)} = -2kdt \tag{2.61}$$

Using the method of partial fractions, which we have already reviewed above, we can rewrite this as

$$\frac{d[B]}{\left(\frac{[A]_0}{2} - [B]_0\right)[B]} - \frac{d[B]}{\left(\frac{[A]_0}{2} - [B]_0\right)\left([B] + \frac{[A]_0}{2} - [B]_0\right)} = -2kdt \tag{2.62}$$

or

$$\frac{d[B]}{[B]} - \frac{d[B]}{\left([B] + \frac{[A]_0}{2} - [B]_0\right)} = -2\left(\frac{[A]_0}{2} - [B]_0\right)kdt \tag{2.63}$$

which integrates to,

$$\frac{d[B]}{[B]}-\frac{d[B]}{\left([B]+\frac{[A]_0}{2}-[B]_0\right)}=-2\left(\frac{[A]_0}{2}-[B]_0\right)kdt \qquad (2.64)$$

or

$$\ln\frac{[B]}{[B]_0}-\ln\frac{[B]+\frac{[A]_0}{2}-[B]_0}{[B]_0+\frac{[A]_0}{2}-[B]_0}=-2\left(\frac{[A]_0}{2}-[B]_0\right)kt \qquad (2.65)$$

or

$$\ln\frac{[B]}{[B]_0}-\ln\frac{[B]+\frac{[A]_0}{2}-[B]_0}{[B]_0+\frac{[A]_0}{2}-[B]_0}=-2\left(\frac{[A]_0}{2}-[B]_0\right)kt \qquad (2.66)$$

Eq. (2.66) is messy, but it can be solved for [B] if desired. The point is that this equation looks very different from our previous solution using [P] as a variable (Equation 30), even though according to the stoichiometry,

$$\frac{d[P]}{dt}=-\frac{d[B]}{dt} \qquad (2.67)$$

This equation has the same problem we had before if the reactants are initially present in their stoichiometrically equivalent amounts.

It is an interesting exercise to show algebraically that Equations (2.50) and (2.66) are, indeed, equivalent.

3

Reaction Mechanisms

Elementary Reaction Steps - Molecularity

We have made a big deal out of the fact that you cannot predict the rate law for a chemical reaction from the balanced equation for that reaction. We now introduce a new class of simple reactions, called elementary reaction steps, for which the rate law IS determined by the balanced reaction equation. We also introduce a new term, molecularity, which tells us the number of molecules involved in an elementary reaction step.

We emphasize that the reaction velocity for each elementary reaction step IS determined by the reaction step and, later on, we will string a sequence of elementary reaction steps together to form what is called a mechanism. A reaction mechanism is a detailed (theoretical) description of how we think the chemical reaction proceeds. That is, it describes our thought about which molecule collides with which other molecule to form an intermediate product, which my go on to react with some other species, and so on, to produce the overall reaction. The elementary reaction steps must be balanced (as do all chemical reactions). We can usually tell the difference between an elementary reaction step and a balanced reaction by the fact that the elementary reaction step will have one or more rate constants associated with it, as we will see below.

It is probably easiest to describe the elementary reaction steps and their associated rate laws by just telling you what they are. As we proceed you will become aware of the fact that the rate law is completely determined by the elementary reaction step you write down. (The converse is not true. We will see in the examples below that several different elementary steps may give rise to the same rate law.)

Unimolecular Reaction Steps

The elementary reaction step,

$$\mathrm{A} \xrightarrow{k_1} \mathrm{B} \tag{3.1}$$

is unimolecular because there is only one molecule reacting, that is, molecule "A" is reacting. This unimolecular reaction step implies the rate law,

$$\frac{d[\mathrm{B}]}{dt} = k_1[\mathrm{A}] \tag{3.2}$$

or, equivalently,

$$\frac{d[\mathrm{A}]}{dt} = k_1[\mathrm{A}] \tag{3.3}$$

In words, these elementary reaction steps say that the molecule, A, spontaneously transforms into B at some rate k_1. The algebraic sign in front of k_1 tells whether you are gaining product or losing reactant depending on whether the concentration in the derivative is increasing or decreasing. For example, in Eq. (3.2), [B] is increasing, and in Eq. 3.3, [A] is decreasing. (All of the usual rules of stoichiometry still hold in elementary reaction steps. If you use up some reactant you must gain an equivalent amount of product.)

An elementary reaction step may be reversible or irreversible. Equation 1 is an irreversible unimolecular step. Equation (3.4_ below is a reversible unimolecular step:

$$A \underset{k_{-1}}{\overset{k_1}{\rightleftharpoons}} B \tag{3.4}$$

This reversible unimolecular step implies the following rate laws,

$$\frac{d[B]}{dt} = +k_1[A] - k_2[B] \tag{3.5}$$

and/or

$$\frac{d[A]}{dt} = -k_1[A] + k_2[B] \tag{3.6}$$

(Either one of these may be used, depending on whether we are trying to account for the disappearance of reactant, A, or the appearance of product, B, in our mechanism for a particular reaction.)

A unimolecular reaction step can have more than one product, for example,

$$A \xrightarrow{k_1} B + C \tag{3.7}$$

The unmimolecular process given by Eq. 3.7 implies the same rate law as the reaction in Eq. 3.1, namely, either Eq. (3.2) or Eq. (3.3).

Bimolecular Reaction Steps

There are several varieties of bimolecular steps. For example,

$$A + B \xrightarrow{k_2} C \tag{3.8}$$

implies the rate law,

$$\frac{d[\mathrm{C}]}{dt} = k_2[\mathrm{A}][\mathrm{B}] \tag{3.9}$$

or

$$\frac{d[\mathrm{A}]}{dt} = -k_2[\mathrm{A}][\mathrm{B}] \tag{3.10}$$

and so on. In Eq. (3.8), (3.9) and (3.10) we have given only one product, "C." We would get the same rate laws if there had been two or more products, for example as in,

$$\mathrm{A} + \mathrm{B} \xrightarrow{k_2} \mathrm{C} + \mathrm{D} \tag{3.11}$$

Reversible Bimolecular Steps

The bimolecular reaction

$$\mathrm{A} + \mathrm{B} \underset{k_{-2}}{\overset{k_2}{\rightleftarrows}} \mathrm{C} \tag{3.12}$$

implies the rate law

$$\frac{d[\mathrm{C}]}{dt} = +k_2[\mathrm{A}][\mathrm{B}] - k_{-2}[\mathrm{C}] \tag{3.13}$$

or it could be written as a rate of loss of A or B as we have seen above. The reversible bimolecular reaction,

$$\mathrm{A} + \mathrm{B} \underset{k_{-2}}{\overset{k_2}{\rightleftarrows}} \mathrm{C} + \mathrm{D} \tag{3.14}$$

implies

$$\frac{d[\mathrm{C}]}{dt} = +k_2[\mathrm{A}][\mathrm{B}] - k_{-2}[\mathrm{C}][\mathrm{D}] \tag{3.15}$$

and its variants.

Termolecular Reaction Steps

Termolecular reaction steps require three molecules coming together at the same time. They are rare because three-body collisions in the gas phase are rare, but there are cases of termolecular reactions in the literature. There are many varieties of termolecular reactions and they may be reversible. A typical termolecular process might be,

$$A + B + C \xrightarrow{k_3} D \tag{3.16}$$

with its implied rate law

$$\frac{d[D]}{dt} = +k_3[A][B][C] \tag{3.17}$$

A reaction mechanism is a combination of one or more elementary reaction steps which start with the appropriate reactants and end with the appropriate product(s). We will illustrate the construction of a reaction mechanism with the famous Lindemann mechanism for a class of gas-phase reactions

The Lindemann Mechanism

This mechanism is sometimes called the Lindemann-Hinshelwood mechanism. Lindemann found it and Hinshelwood polished it. There is a class of gas-phase reactions of which one example is:

$$N_2\,O_5 = NO_2 + NO_3 \tag{3.18}$$

In chemical kinetics experiments sometimes this reaction looks like a first order reaction and sometimes it looks like a second order reaction, depending on the reaction conditions. For example, you might change the apparent order of the reaction by changing the pressure.

The postulated mechanism is:

$$N_2O_5 + N_2O_5 \underset{k_{-2}}{\overset{k_2}{\rightleftharpoons}} N_2O_5^* + N_2O_5 \quad (3.19)$$

$$N_2O_5^* \xrightarrow{k_1} NO_2 + NO_3 \quad (3.20)$$

The $N_2O_5^*$ is said to be "activated" N_2O_5. The rate fo the reaction is given by the rate at which the activated N_2O_5 falls apart into products. We can see from Equation 3 that the formation of products is a unimolecular process which has a rate law of

$$\frac{d[NO_2]}{dt} = +k_1[N_2O_5^*] \quad (3.21)$$

The activated N_2O_5 (that is, the $N_2O_5^*$) is a transient species because it is neither a reactant nor a product. It is produced during the course of the reaction and it is gone at the end of the reaction.

We can also write a rate law for the change in concentration of this transient species. We do this by looking at all of the elementary steps in the proposed mechanism which produce or destroy this transient species. Notice that from Eq. (3.19) the transient species is produced by the forward process with rate constant, k_2, and it is destroyed by the reverse process with rate constant, k_{-2}. The transient species is also destroyed by the process in Eq. (3.20), where it "falls apart" to give the products at a rate, k_1. So the elementary reaction steps that produce the transient species must appear in the rate equation for the transient with a positive sign and the steps that destroy the transient appear with a negative sign. Thus the complete description for the rate of change of the concentration of the transient is,

$$\frac{d[N_2O_5^*]}{dt} = k_2[N_2O_5]^2 - k_{-2}[N_2O_5][N_2O_5^*] - k_1[N_2O_5^*] \quad (3.22)$$

Eqs. (3.21) and (3.22) form a set of coupled first order differential equations in time which can be solved exactly, but it is easier to make an approximation which will allow us to obtain a simple, but approximate, solution.

This approximation is called the steady state approximation and it will allow us to obtain, relatively easily, a simple solution which is good enough for our purposes. It turns out that the concentrations of transient species usually build up quickly and decay away very slowly during the course of the reaction, and the concentrations of transient species usually do not ever get very large. If this is the case (and we will assume that it is the case in this course) we can say that the rate of change of the concentration of transient species is essentially zero over most of the course of the reaction.

Our procedure is then to look for transient species and approximate their rate of change as small (say zero) over most of the course of the reaction. That is, we set

$$\frac{d[N_2O_5^*]}{dt} \approx 0 \tag{3.23}$$

Then Eq. (3.22) becomes,

$$k_2[N_2O_5]^2 - k_{-2}[N_2O_5][N_2O_5^*] - k_1[N_2O_5^*] = 0 \tag{3.24}$$

Solve this for $[N_2O_5^*]$ to get,

$$[N_2O_5^*] = \frac{k_2[N_2O_5]^2}{k_1 + k_{-2}[N_2O_5]} \tag{3.25}$$

Then we can insert this transient species concentration back into our original rate equation for the formation of product (Eq. 3.21) to obtain an overall rate equation for the reaction,

$$\text{Rate} = k_1[N_2O_5^*] = \frac{k_1 k_2[N_2O_5]^2}{k_1 + k_{-2}[N_2O_5]} \tag{3.26}$$

We now must ask whether or not this rate law fits the experimental fact that sometimes the reaction appears to be first order in N_2O_5 and sometimes it appears to be second order in N_2O_5.

To check this we look at Eq. (3.26) in two limiting cases:

If $k_1 >> k_{-2}$ $[N_2O_5]$, that is, when $[N_2O_5]$ is small, then the

$$\text{Rate} \sim k_2[N_2O_5]^2 \tag{3.27}$$

and the reaction looks like it is second order.

If $k_1 << k_{-2}$ $[N_2O_5]$, that is, when $[N_2O_5]$ is large, then the

$$\text{Rate} \approx \frac{k_1 k_2}{k_{-2}}[N_2O_5] \tag{3.28}$$

and the reaction looks like it is first order.

There are lots of variations to the Lindemann mechanism. (From now on we will use the letters A, B, C, ..., etc. to designate molecules.)

Let's look at a variation where there is an inert gas, M, present. The proposed mechanism is,

$$A + A \underset{k_{-2}}{\overset{k_2}{\rightleftharpoons}} A^* + A \tag{3.29}$$

$$A + M \underset{k'_{-2}}{\overset{k'_2}{\rightleftharpoons}} A^* + M \tag{3.30}$$

$$A^* \xrightarrow{k_1} B + C \tag{3.31}$$

The rate is still given by an equation similar to Eq. 3.21

$$\frac{d[B]}{dt} = k_1[A^*] \tag{3.32}$$

But for the transient species we get

$$\frac{d[A^*]}{dt} = k_2[A]^2 - k_{-2}[A][A^*] + k'_2[A][M] - k'_{-2}[A^*][M] - k_1[A^*] \approx 0 \quad (3.33)$$

In Eq. (3.33) we have already applied the steady state approximation by setting this rate as approximately equal to zero. Using the steady state approximation we solve for [A*],

$$[A^*] = \frac{k_2[A]^2 + k'_2[A][M]}{k_1 + k_{-2}[A] + k'_{-2}[M]} \quad (3.34)$$

and plug this into Eq. (3.32) to get the reaction rate as,

$$\text{Rate} = k_1[A^*] = k_1 \frac{k_2[A]^2 + k'_2[A][M]}{k_1 + k_{-2}[A] + k'_{-2}[M]} \quad (3.35)$$

If [M] = 0 then we are back to the Lindemann mechanism. However, there are new possibilities. We can change the appearance of the kinetics by manipulating [M]. For example, if we make [M] very large, large enough to overwhelm all the other terms in the rate law, the effective rate law becomes,

$$\text{Rate} \to k_1 \frac{k'_2[A][M]}{k'_{-2}[M]} = \frac{k_1 k'_2}{k'_{-2}}[A] \quad (3.36)$$

That is, the reaction becomes effectively first order in A independent of the values of either [A] or [M] (as long as [M] is large).

Sometimes the conditions are such that there is virtually no "self activation" of the A molecules and all of the activation comes from collisions of A molecules with the inert gas molecules, M. Our mechanism covers this case. We can obtain the rate law in this case by setting $k2$ and k-2 equal to zero. This gives

$$\text{Rate} \rightarrow k_1 \frac{k'_2[\text{A}][\text{M}]}{k'_{-2}[\text{M}]} = \frac{k_1 k'_2}{k'_{-2}}[\text{A}] \quad (3.37)$$

In this situation the reaction is first order in A regardless of the M concentration, but the actual reaction rate depends on [M]. One can define an effective first order rate constant by rewriting the rate equation as

$$\frac{d[\text{B}]}{dt} = k_1 \frac{k'_2[\text{M}]}{k_1 + k'_{-2}[\text{M}]}[\text{A}] \quad (3.38)$$

so that the effective rate constant is

$$k_{\text{eff}} = k_1 \frac{k'_2[\text{M}]}{k_1 + k'_{-2}[\text{M}]} \quad (3.39)$$

Clearly you can change the effective first order rate constant, and hence the half-life of the reaction, by adjusting [M]. These examples illustrate a statement that was made earlier in our study of chemical kinetics, namely that the rate law may contain species that are not part of the balanced chemical reaction. Other examples of species not in the balanced reaction occuring in the rate law would include catalysis, where a catalyst does not normally appear in the balanced reaction but does appear in the rate law.

Chain Reactions

Chain reactions usually involve free radicals. We will show how to deal with chain reactions by working out several examples. Our first example is the gas phase reaction of hydrogen with bromine to give HBr. (The temperature must be high enough that bromine is a gas and not a liquid.)

$$H_2 + Br_2 = 2\ HBr \quad (3.40)$$

The experimental rate law is

$$\frac{d[\text{HBr}]}{dt} = \frac{k'[\text{H}_2][Br_2]^{1/2}}{1+k''\frac{[\text{HB}r]}{[\text{Br}_2]}}[\text{A}] \qquad (3.41)$$

Our task is to show that the postulated mechanism will give this rate law. (We will not try to invent mechanisms for particular reactions in this course.) The postulated mechanism has five elementary reaction steps:

Initiation	$Br_2 \xrightarrow{ka} Br\bullet + Br\bullet$	(3.42)
Propagation	$BR\bullet + H_2 \xrightarrow{kb} HBr + H\bullet$	(3.43)
Propagation	$H\bullet + Br_2 \xrightarrow{kc} HBr + Br\bullet$	(3.44)
Inhibition	$H\bullet + HBr \xrightarrow{kd} H_2 + Br\bullet$	(3.45)
Termination	$Br\bullet + Br\bullet \xrightarrow{ke} Br_2$	(3.46)

Our goal is to find the rate law implied by this mechanism and see how it comparaes to the experimental rate law, Equation 2. Our first task will be to define the reaction rate. We will define the reaction rate as rate of production of the product, HBr.

$$\text{Rate} = \frac{d[\text{HBr}]}{dt} \qquad (3.47)$$

Looking at the mechanism and using only the steps which yield or consume HBr we find that

$$\frac{d[\text{HBr}]}{dt} = k_b[\text{Br}\bullet][\text{H}_2] + k_c[\text{H}\bullet][\text{Br}_2] - k_d[\text{H}\bullet][\text{HBr}] \qquad (3.48)$$

We see that the rate contains two transient species, Br• and H• , so we set up the equations which show the rate of change of the concentrations of these two species and apply the steady state approximation. The two equations are,

$$\frac{d[\mathrm{H\bullet}]}{dt} = k_\mathrm{b}[\mathrm{Br\bullet}][\mathrm{H_2}] - k_\mathrm{c}[\mathrm{H\bullet}][\mathrm{Br_2}] - k_\mathrm{d}[\mathrm{H\bullet}][\mathrm{HBr}] \approx 0 \quad (3.49)$$

and

$$\frac{d[\mathrm{Br\bullet}]}{dt} = 2k_\mathrm{a}[\mathrm{Br_2}] + \{-k_\mathrm{b}[\mathrm{Br\bullet}][\mathrm{H_2}] + k_\mathrm{c}[\mathrm{H\bullet}][\mathrm{Br_2}] + k_\mathrm{d}[\mathrm{H\bullet}][\mathrm{HBr}]\} - 2k_\mathrm{e}[\mathrm{Br\bullet}]^2 \approx 0 \quad (3.50)$$

In Eq. (3.50) we have set three of the terms apart by braces { } in order to make it easier to see that the part inside the braces is just the negative of the middle of Eq. (3.49). Since the part inside the braces is zero we conclude that

$$2k_a\,[\mathrm{Br_2}] - 2k_e\,[\mathrm{Br\bullet}]^2 = 0 \quad (3.51)$$

or

$$[\mathrm{Br\bullet}] = \sqrt{\frac{k_\mathrm{a}}{k_\mathrm{e}}[\mathrm{Br_2}]} \quad (3.52)$$

We can solve Eq. (3.49) for [H•] to get

$$[\mathrm{H\bullet}] = \frac{k_b[\mathrm{Br\bullet}][\mathrm{H_2}]}{k_\mathrm{c}[\mathrm{Br_2}] + k_d[\mathrm{HBr}]} \quad (3.53)$$

which, on substituting the result of Eq. (3.52), becomes,

$$[\mathrm{H\bullet}] = \frac{k_b[\mathrm{H_2}]\sqrt{\frac{k_a}{k_e}[\mathrm{Br_2}]}}{k_\mathrm{c}[\mathrm{Br_2}] + k_d[\mathrm{HBr}]} \quad (3.54)$$

We now go back to the equation for the Rate, Eq. (3.48), and note that the right hand side of Eq. (3.48) is the same as Eq. (3.49) except for the sign of the kc term. Rewrite Eq. (3.49) as,

$$k_b[\mathrm{Br\bullet}][\mathrm{H_2}] - k_c[\mathrm{H\bullet}][\mathrm{Br_2}] - k_d\,[\mathrm{H\bullet}][\mathrm{HBr}] = 0 \quad (3.55)$$

or

$$k_b[\text{Br}\bullet][\text{H}_2] - k_d[\text{H}\bullet][\text{HBr}] = k_c\,[\text{H}\bullet][\text{Br}_2] \quad (3.56)$$

Combining Eq. (3.48) and (3.56) we find that,

$$\frac{d[HBr]}{dt} = 2k_c[\text{H}\bullet][\text{Br}_2] \quad (3.57)$$

Since Eq. (3.54) gives us an expression for [H•] we can combine Eq. (3.54) and (3.55) to eliminate all the transient concentrations,

$$\frac{d[HBr]}{dt} = 2k_c\,\frac{k_b[\text{H}_2]\sqrt{\frac{k_a}{k_e}[\text{Br}_2]}}{k_c[\text{Br}_2]+k_d[\text{HBr}]}[\text{Br}_2] \quad (3.58)$$

which rearranges to

$$\frac{d[\text{HBr}]}{dt} = \frac{2k_b\sqrt{\frac{k_a}{k_e}}[\text{H}_2][\text{Br}_2]^{1/2}}{1+\frac{k_d}{k_c}\frac{[\text{HBr}]}{[\text{Br}_2]}} \quad (3.59)$$

Eq. (3.59) has the same form as Equation 2 with,

$$k' = 2k_b\sqrt{\frac{k_a}{k_e}} \quad (3.60)$$

and

$$k'' = \frac{k_d}{k_c} \quad (3.61)$$

The two phenomenological rate constants, k' and k'', can be written in terms of the rate constants for the elementary reaction steps in the mechanism.

Another Example of a Chain Reaction

The reaction (gas phase)

$$C_2H_6 = C_2H_4 + H_2 \quad (3.62)$$

is a very important commercial reaction because ethylene (ethene) is a precursor to many important products (of which polyethylene is just one). Because of its commercial importance the reaction has been much studied. In the temperature range 700 K to 900 K the early stages of the reaction appear to be governed by the mechanism,

$$\text{Initiation} \quad C_2H_6 \xrightarrow{k_1} 2\ CH_3^{\bullet} \quad (3.63)$$

$$\text{Transfer} \quad CH_3^{\bullet} + C_2H_6 \xrightarrow{k_2} CH_4 + C_2H_5^{\bullet} \quad (3.64)$$

$$\text{Propagation} \quad C_2H_5 \xrightarrow{k_3} C_2H_4 + H\bullet \quad (3.65)$$

$$\text{Propagation} \quad H\bullet + C_2H_6 \xrightarrow{k_4} C_2H_5\bullet + H_2 \quad (3.66)$$

$$\text{Termination} \quad H\bullet + C_2H_5 \xrightarrow{k_5} C_2H_6 \quad (3.67)$$

Define the rate as the rate of production of C_2H_4 from Eq. (3.65),

$$\text{Rate} = \frac{d[C_2H_4]}{dt} = k_3[C_2H_5\bullet] \quad (3.68)$$

There are three transient species, $CH_3\bullet$, $C_2H_5\bullet$, and $H\bullet$. We will use the steady state approximation on each of these:

$$\frac{d[CH_3\bullet]}{dt} = 2k_1[C_2H_6] - k_2[CH_3\bullet][C_2H_6] \approx 0 \quad (3.69)$$

$$\frac{d[C_2H_5\bullet]}{dt} = k_2[CH_3\bullet][C_2H_6] - k_3[C_2H_5\bullet] + k_4[H\bullet][C_2H_6] - k_5[H\bullet][C_2H_5\bullet] \approx 0 \quad (3.70)$$

$$\frac{d[\mathrm{H}\bullet]}{dt} = k_3[\mathrm{C_2H_5}\bullet] - k_4[\mathrm{H}\bullet][\mathrm{C_2H_6}] - k_5[\mathrm{H}\bullet][\mathrm{C_2H_5}\bullet] \approx 0 \quad (3.71)$$

From Eq. (3.69) we conclude that

$$[\mathrm{CH_3}\bullet] = \frac{2k_1}{k_2} \quad (3.72)$$

Equation (3.72), says that the concentration of the methyl radical remains constant throughout the course of the reaction. This, of course, is impossible because the concentration of the methyl must be zero at the beginning and end of the reaction. However, the steady state approximation is still useful if the concentration builds up to this constant concentration fairly quickly, remains essentially constant throughout the portion of the reaction for which this mechanism is valid.)

If we add Eq. (3.70) and (3.71) and substitute in the result of Eq. (3.72) we get,

$$k_2\frac{2k_1}{k_2}[\mathrm{C_2H_6}] - 2k_5[\mathrm{H}\bullet][\mathrm{C_2H_5}\bullet] = 0 \quad (3.73)$$

which simplifies to

$$[\mathrm{C_2H_5}\bullet] = \frac{k_1}{k_5}\frac{1}{[\mathrm{H}\bullet]}[\mathrm{C_2H_6}] \quad (3.74)$$

We can get another expression for $[\mathrm{C_2H_5}\bullet]$ from Eq. (3.70)

$$[\mathrm{C_2H_5}\bullet] = \frac{k_2\frac{2k_1}{k_2}+k_4[\mathrm{H}\bullet]}{k_3+k_5[\mathrm{H}\bullet]}[\mathrm{C_2H_6}] \quad (3.75)$$

When we equate Eqs. (3.74) and (3.75) and simplify we get

$$\frac{k_1}{k_5}\frac{1}{[\mathrm{H}\bullet]} = \frac{2k_1+k_4[H\bullet]}{k_3+k_5[H\bullet]} \tag{3.76}$$

Equation (3.70) is clearly a quadratic in [H•], but the coefficients (in other words, everything else in the equation) are all constants so [H•] must equal a constant, call it k',

$$[\mathrm{H}\bullet] = k'$$

Then

$$[\mathrm{C_2H_5}\bullet] = \frac{2k_1+k_4k'}{k_3+k_5k'}[\mathrm{C_2H_6}] \tag{3.77}$$

and the rate becomes, combining Eqs. (3.68) and (3.77),

$$\text{Rate} = \mathrm{k}_3\ \frac{2k_1+k_4k'}{k_3+k_5k'}[\mathrm{C_2H_6}] \tag{3.78}$$

which is first order in ethane. That is, in the early stages of the cracking of ethane to ethylene (ethene) and hydrogen the reaction is first order in ethane.

(If you want to get picky, the above quadratic can be solved to give,

$$[\mathrm{H}\bullet] = \frac{-k_1k_5 \pm \sqrt{k_1k_5(k_1k_5+4k_3k_4)}}{2k_4k_5} = k' \tag{3.79}$$

where we will have to pick the plus sign because concentrations cannot be negative.)

4

Enzyme Kinetics

Enzymes are nature's catalysts. The vast majority of chemical reactions that keep living things alive are much too slow (without a catalyst) to sustain life. An example of this is the oxidation of a sugar - say glucose - to give water, carbon dioxide and energy. You can leave glucose open to the air for years without any appreciable oxidation, yet this is one of the reactions that provides the energy to walk and run in daily life. There are diseases caused by the failure of the body to produce a specific enzyme. The Michaelis-Menten mechanism for the catalysis of biological chemical reactions is one of the most important chemical reaction mechanisms in biochemistry.

The Michaelis-Menten mechanism for enzyme kinetics is:

$$E + S \underset{k_{-1}}{\overset{k_1}{\rightleftarrows}} ES \xrightarrow{k_2} E + \text{products} \tag{4.1}$$

E is the enzyme, S is the "substrate" (the molecule on which the enzyme does its work), and ES is an enzyme-substrate complex. (It is presumed that the substrate must somehow bind to the enzyme before the enzyme can do its work.)

We analyze this mechanism as usual. First, we define the reaction rate as the rate of formation of product and write the kinetic equation implied by this mechanism,

$$\frac{d[\text{products}]}{dt} = k_2[\text{ES}] \tag{4.2}$$

The enzyme-substrate complex, ES, is a transient species so we set up an equation for its rate of change and apply the steady state approximation,

$$\frac{d[\text{ES}]}{dt} = k_1[\text{E}][\text{S}] - k_{-1}[\text{ES}] - k_2[\text{ES}] \approx 0 \tag{4.3}$$

Solve for [ES]

$$[\text{ES}]\frac{k_1[\text{E}][\text{S}]}{k_{-1} + k_2} \tag{4.4}$$

and substitute it into the equation for the rate,

$$\text{Rate} = \frac{d[\text{product}]}{dt} = k_2 \frac{k_1[\text{E}][\text{S}]}{k_{-1}+k_2} \tag{4.5}$$

We might think that we are finished, but there is a complication and some new notation to introduce. First we introduce the Michaelis-Menten constant, K_M,

$$\frac{1}{K_M} = k_2 \frac{k_1}{k_{-1}+k_2} \tag{4.6}$$

so that the rate becomes,

$$\text{Rate} = \frac{1}{K_{\text{M}}} = k_2 \frac{k_2}{K_M}[\text{E}][\text{S}] \tag{4.7}$$

Now we must deal with the difficulty that [E] is the concentration of free (uncomplexed) enzyme and this is usually not known. What is known is the total enzyme concentration, $[\text{E}]_o$, but

$$[\text{E}]_o = [\text{E}] + [\text{ES}] = [\text{E}] + \frac{[\text{E}][\text{S}]}{K_{\text{M}}}$$
$$= [\text{E}]\left(\frac{[\text{S}]}{K_{\text{M}}}\right) \tag{4.8}$$

for which we obtain

$$[\text{E}] = \frac{[\text{E}]_o}{\left(1 + \frac{[\text{S}]}{K_M}\right)} \tag{4.9}$$

The rate becomes, then,

$$\text{Rate} = \frac{k_2}{K_{\text{M}}}[\text{S}]\frac{[\text{E}]_0}{\left(1+\frac{[\text{S}]}{K_{\text{M}}}\right)}$$
$$= k_2 \frac{[\text{E}]_0[\text{S}]}{K_{\text{M}}+[\text{S}]} \tag{4.10}$$

Define the reaction velocity as v = Rate. So,

$$v = k_2 \frac{[\text{E}]_0[\text{S}]}{K_{\text{M}}+[\text{S}]} \tag{4.11}$$

Note that the reaction velocity, v, is zero when [S] is zero and that the reaction velocity increases as we increase [S]. The reaction velocity reaches a maximum when [S] becomes very large. Define the maximum velocity, v_{max}, as,

$$v_{\max} = \lim_{[S]\to\infty} k_2 \frac{[E]_0[S]}{K_M+[S]} = k_2[E]_0 \tag{4.12}$$

then

$$v = \frac{v_{\max}[S]}{K_M+[S]} \tag{4.13}$$

Note that the kinetics of the reaction are characterized by two parameters, v_{max} and K_M. These are the parameters that are usually given in the literature in studies of the kinetics of biochemical reactions.

In order to deal with experimental data we write,

$$\frac{1}{v} = \frac{K_M}{v_{\max}[S]} + \frac{1}{v_{\max}} \tag{4.14}$$

In an experiment one measures v as a function of [S]. If we plot $1/v$ against $1/[S]$ we should get a straight line with slope, K_M/v_{max} and intercept $1/v_{max}$. This gives us both parameters,

$$K_M = \text{Slope} \times v_{\max} = \frac{\text{Slope}}{\text{Intercept}} \tag{4.15}$$

Enzyme with Inhibitor

Recall the basic Michaelis-Menten mechanism,

$$E + S \underset{k_{-1}}{\overset{k_1}{\rightleftarrows}} ES \xrightarrow{k_2} E + \text{products} \tag{4.1}$$

There are several possibilities for an inhibitor, I, to interfere with this reaction:

(1) $$E + I \underset{k'_{-1}}{\overset{k'_1}{\rightleftarrows}} EI \tag{4.16}$$

In words, the inhibitor binds with the enzyme to the exclusion of the substrate.

$$\text{(2)} \qquad ES + I \underset{k''_{-1}}{\overset{k''_1}{\rightleftharpoons}} ESI \tag{4.17}$$

In words, the inhibitor binds to the enzyme-substrate complex and alters the action of the enzyme on the substrate. You can have (1) or (2) or both. We will only work out the first case. The procedure is the same as for the uninhibited Michaelis-Menten mechanism except for an additional term in the expression for the total enzyme concentration and a new transient, EI. The rate is still

$$\frac{d[\text{products}]}{dt} = k_2\,[ES] \tag{4.18}$$

and we apply the steady state approximation to ES, which leads to

$$[ES] = \frac{k_1[E][S]}{k_{-1} + k_2} \tag{4.19}$$

and the same rate expression

$$\text{Rate} = \frac{d[\text{product}]}{dt} = k_2 \frac{k_1[E][S]}{k_{-1}+k_2} \tag{4.20}$$

Use the same definition of K_M,

$$\frac{1}{K_M} = \frac{k_1}{k_{-1}+k_2} \tag{4.21}$$

which leads to,

$$\text{Rate} = \frac{k_2}{K_M}[E][S] \tag{4.22}$$

But the enzyme-inhibitor complex is also a transient,

$$\frac{d[\mathrm{EI}]}{dt} = k'_1[\mathrm{E}][\mathrm{I}] - k'_{-1}[\mathrm{EI}] \approx 0 \tag{4.23}$$

giving

$$[\mathrm{EI}] = \frac{k'_1}{k'_{-1}}[\mathrm{E}][\mathrm{I}] \tag{4.24}$$

Now the total enzyme concentration has an extra term

$$[\mathrm{E}]_o = [\mathrm{E}][\mathrm{ES}] + [\mathrm{EI}] = [\mathrm{E}] + \frac{[\mathrm{E}][\mathrm{S}]}{K_\mathrm{M}} + \frac{k'_1}{k'_{-1}}[\mathrm{E}][\mathrm{I}] \tag{4.25a}$$

$$= [\mathrm{E}]\left(1 + \frac{[\mathrm{S}]}{K_\mathrm{M}} + \frac{k'_1}{k'_{-1}}\right)[\mathrm{I}] \tag{4.25b}$$

leading to

$$[\mathrm{E}] = \frac{[\mathrm{E}]_o}{\left(1 + \frac{[\mathrm{S}]}{K_\mathrm{M}} + \frac{k'_1}{k'_{-1}}[\mathrm{I}]\right)} \tag{4.26}$$

and the rate is

$$\text{Rate} = \frac{k_2}{K_\mathrm{M}}[\mathrm{S}]\frac{[\mathrm{E}]_o}{\left(1 + \frac{[\mathrm{S}]}{K_\mathrm{M}} + \frac{k'_1}{k'_{-1}}[\mathrm{I}]\right)} \tag{4.27}$$

v_{mas} is still

$$v_{max} = k_2[\mathrm{E}]_o \tag{4.28}$$

so the reaction velocity becomes

$$v = \frac{v_{max}[S]}{K_M + [S] + K_M \frac{k'_1}{k'_{-1}}[I]} \qquad (4.29)$$

and then

$$\frac{1}{v} = \frac{K_M}{v_{max}[S]} + \frac{1}{v_{max}} + \frac{K_M}{v_{max}} + \frac{k'_1}{k'_{-1}}\frac{[I]}{[S]} \qquad (4.30)$$

We still plot $1/v$ against $1/[S]$ and the intercept is $1/v_{max}$, but the slope is

$$\text{slope} = \frac{K_M}{v_{max}} + \frac{K_M}{v_{max}}\frac{k'_1}{k'_{-1}}[I] \qquad (4.31)$$

and

$$\frac{\text{slope}}{\text{intercept}} = K_M\left(1 + \frac{k'_1}{k'_{-1}}\right)[I] \qquad (4.32)$$

We must do several different experiments at different [I] to get K_M and k'_1/k'_{-1}. Note that you can only get the ratio k'_1/k'_{-1} and not the individual k's.

Relaxation Methods

Manfred Eigen received the Nobel prize for this work in 1967.

The so-called "relaxation methods" are methods for finding the rate constants of very fast reactions, for example, consider the reaction in liquid water,

$$H^+ + OH^- \rightarrow H_2O \qquad (4.33)$$

This reaction is so fast that it looks instantaneous. If you add NaOH(*aq*) to HCl(*aq*) the reaction is complete as soon as you get it mixed. Another example reaction, which you may have done in the laboratory, is the formation of barium sulfate,

$$Ba^{2+}(aq) + SO_4^{2-}(aq) = BaSO_4(s) \tag{4.34}$$

When you pour a barium chloride solution into dilute sulfuric acid the barium sulfate precipitate forms so fast that you can't tell that it is not instantaneous. (Chemical reactions are not instantaneous, as we shall see, but they can be very fast.)

The idea of a relaxation method is to disturb an existing equilibrium and watch it decay back to equilibrium. You can disturb the equilibrium by a sudden small change in temperature (called the T-jump method), or by a sudden small increase in pressure (called the p-jump method), or in some other way. You can follow the progress of the reaction as it returns to equilibrium electronically by measuring the change in some property like electrical conductivity or spectroscopic absorption, or etc.

For example, consider the reaction,

$$H + OH^- \underset{k_1}{\overset{k_2}{\rightleftarrows}} H_2O \tag{4.35}$$

(This is really the reaction

$$H_3O^+ + OH^- \underset{k_{-2}}{\overset{k_2}{\rightleftarrows}} 2H_2O \tag{4.36}$$

but we will ignore that complication)

The rate of formation of water is,

$$\frac{d[H_2O]}{dt} = k2[H^+][OH^-] - k_1[H_2O] \tag{4.37}$$

At equilibrium the net rate of formation of water is zero so

$$\frac{d[H_2O]}{dt} = k_2[H^+]_{eq}[OH^-]_{eq} - k_1[H_2O]_{eq} = 0 \tag{4.38}$$

Disturb the equilibrium by a small amount. (This can be done by giving the system a very fast pulse of heat or a sudden increase in pressure, etc.) After the disturbance,

$$[H]^+ = [H^+]_{eq} + \delta \tag{4.38a}$$

$$[OH^-] = [OH^-]_{eq} + \delta \tag{4.38b}$$

$$[H_2O] = [H_2O]_{eq} - \delta \tag{4.38c}$$

(Note that the stoichiometry is preserved.) Then, in terms of *d*,

$$\frac{d[H_2O]}{dt} = -\frac{d\delta}{dt} = k_2\left([H^+]_{eq} + \delta\right)\left([OH^-]_{eq} + \delta\right)$$

$$I - k_1([H_2O]_{eq} - \delta) \tag{4.39}$$

which can be expanded to

$$-\frac{d\delta}{dt} = k_2 [H^+]_{eq} [OH^-]_{eq} - k_1 [H_2O]_{eq} + k_2 \delta ([H^+]_{eq} \tag{4.40}$$

The first two terms on the right vanish because they are the equilibrium rate of production of water. If *d* is small then d^2 can be neglected in comparison to it. Our rate equation becomes

$$-\frac{d\delta}{dt} = \delta (k_2[H^+]_{eq} + k_2[OH^-]_{eq} + k_1) \tag{4.41}$$

This is a first order process with effective rate constant

$$k_{eff} = \frac{1}{\tau} = k_2[H^+]_{eq} + k_2[OH^-]_{eq} + k_1 \tag{4.42}$$

write

$$\frac{d\delta}{dt} = -\,k_{eff}\,\delta \tag{4.43}$$

and

$$\delta = \delta_o e^{-k_{eff}+} = \delta_o e^{\frac{-t}{\tau}} \tag{4.44}$$

where

$$\tau = \frac{1}{k_2[H^+]_{eq} + k_2[OH^-]_{eq} + k_1} \tag{4.45}$$

t is called the 'relaxation time."

You can adjust the H+ and OH- concentrations by adjusting the pH. Run several experiments at different pH and measure t for each case. For example, experiment 1, at equilibrium, gives,

$$\frac{1}{\tau} = k_2[H^+]_1 + k_2[OH^-]_1 + k_1 \tag{4.46}$$

and experiment 2, at equilibrium, gives

$$\frac{1}{\tau} = k_2[H^+]_2 + k_2[OH^-]_2 + k_1 \tag{4.47}$$

We can solve for k1 and k2. In this case, for example,

$$k_2 = \frac{\frac{1}{\tau_2}\frac{1}{\tau_1}}{[H^+]_2 - [H^+]_1 + [OH^-]_2 - [OH^-]_1} \tag{4.48}$$

The Eigen method has been used to study and find the rates of a great number of very fast chemical reactions. Two examples are,

$H^+ OH^- \rightarrow H_2O, k_2 = 1.4 \times 10^{11} L/mol\ s, k_1 = 2.5 \times 10^{-5}\ s^{-1}$ (4.49a)

$D^+ OD^- \rightarrow D_2O, k_2 = 8.4 \times 10^{10}\ L/mol\ s, k_1 = 2.5 \times 10^{-6}\ s^{-1}$ (4.49b)

Notice that when we substitute deuterium for hydrogen the reaction is slower. That is because the heavier deuterium atom moves slower than a hydrogen atom at the same temperature.

Here is something interesting. We already know that at equilibrium,

$$0 = k_2\,[H^+]_{eq}\,[OH^-]_{eq} - [H_2O]_{eq} \tag{4.50}$$

so

$$[H^+]_{eq}\,[OH^-]_{eq} = \frac{k_1}{k_2}[H_2O]_{eq} \approx \frac{k_1}{k_2} 56 \tag{4.51}$$

$$= 1.0 \times 10^{-14}$$

The 1.0×10^{-14} should look familiar as K_W, the ion product of water.

This illustrates the principle of detailed balance. The principle of detailed balance simply says that at equilibrium the forward and reverse reaction rates in an elementary reaction step must be equal. For a reaction

$$A + B \underset{k_R}{\overset{k_F}{\rightleftarrows}} C + D \tag{4.52}$$

The rate of production of C is given by

$$\frac{d[C]}{dt} = k_F[A][B] - k_r[C][D] \tag{4.53}$$

so that at equilibrium

$$\frac{d[C]}{dt_{eq}} = k_F[A]_{eq}[B]_{eq} - k_R\,[C]_{eq}\,[D]_{eq} = 0 \tag{4.54}$$

We can rearrange this to show that

$$\frac{[\mathrm{C}]_{\mathrm{eq}}[D]_{\mathrm{eq}}}{[\mathrm{A}]_{\mathrm{eq}}[\mathrm{B}]_{\mathrm{eq}}} = \frac{k_{\mathrm{F}}}{k_{\mathrm{R}}} = K_{\mathrm{eq}} \tag{4.55}$$

If a reaction consists of several elementary steps this principle applies to each of the steps individually. The principle of detailed balance is a consequence of "microscopic reversibility" which simply states that the fundamental laws of physics are the same for time going either forward or backward. (A fancy way to say this is to say that the laws of physics are invariant under time reversal.)

5

Chemical Reaction

A chemical reaction is a process that always results in the interconversion of chemical substances The substance or substances initially involved in a chemical reaction are called reactants. Chemical reactions are usually characterized by a chemical change, and they yield one or more products, which usually have properties different from the reactants. Classically, chemical reactions encompass changes that strictly involve the motion of electrons in the forming and breaking of chemical bonds, although the general concept of a chemical reaction, in particular the notion of a chemical equation, is applicable to transformations of elementary particles, as well as nuclear reactions.

Different chemical reactions are used in combination in chemical synthesis in order to get a desired product. In biochemistry, series of chemical reactions catalyzed by enzymes form metabolic pathways, by which syntheses and decompositions ordinarily impossible in conditions within a cell are performed.

Reaction Types

The large diversity of chemical reactions and approaches to their study results in the existence of several concurring, often overlapping, ways of classifying them. Below are examples of widely used terms for describing common kinds of reactions.

- **Isomerisation,** in which a chemical compound undergoes a structural rearrangement without any change in its net atomic composition.
- **Direct combination or synthesis,** in which 2 or more chemical elements or compounds unite to form a more complex product:

  $N_2 + 3\ H_2 \rightarrow 2\ NH_3$
- **Chemical decomposition or analysis,** in which a compound is decomposed into smaller compounds or elements:

  $2\ H_2O \rightarrow 2\ H_2 + O_2$
- **Single displacement or substitution,** characterized by an element being displaced out of a compound by a more reactive element:

  $2\ Na(s) + 2\ HCl(aq) \rightarrow 2\ NaCl(aq) + H_2(g)$
- **Metathesis or Double displacement reaction,** in which two compounds exchange ions or bonds to form different compounds:

  $NaCl(aq) + AgNO_3\ (aq) \rightarrow NaNO_3\ (aq) + AgCl(s)$
- **Acid-base reactions,** broadly characterized as reactions between an acid and a base, can have different definitions depending on the acid-base concept employed. Some of the most common are:
  - *Arrhenius definition:* Acids dissociate in water releasing H_3O^+ ions; bases dissociate in water releasing OH^- ions.

- *Brønsted-Lowry definition:* Acids are proton (H^+) donors; bases are proton acceptors. Includes the Arrhenius definition.
- *Lewis definition:* Acids are electron-pair acceptors; bases are electron-pair donors. Includes the Brønsted-Lowry definition.

- **Redox reactions,** in which changes in oxidation numbers of atoms in involved species occur. Those reactions can often be interpreted as transferences of electrons between different molecular sites or species. An example of a redox reaction is:

  $2\ S_2O_3^{\ 2-}(aq) + I_2\ (aq) \rightarrow S_4O_6^{\ 2-}(aq) + 2\ I^-\ (ab)$

  In which I_2 is reduced to I^- and $S_2O_3^{2-}$ (thiosulfate anion) is oxidized to $S_4O_6^{2-}$.

- **Combustion,** a kind of redox reaction in which any combustible substance combines with an oxidizing element, usually oxygen, to generate heat and form oxidized products. The term combustion is usually used for only large-scale oxidation of whole molecules, i.e. a controlled oxidation of a single functional group is not combustion.

  $C_{10}H_8 + 12\ O_2 \rightarrow 10\ CO_2 + 4\ H_2\ O$

  $CH_2S + 6\ F_2 \rightarrow CF_4 + HF + SF_6$

- **Disproportionation** with one reactant forming two distinct products varying in oxidation state.

  $2\ Sn^{2+} \rightarrow Sn + Sn^{4+}$

- **Organic reactions** encompass a wide assortment of reactions involving compounds which have carbon as the main element in their molecular structure. The reactions in which an organic compound may take part are largely defined by its functional groups.

Chemical Kinetics

The rate of a chemical reaction is a measure of how the concentration or pressure of the involved substances changes with time. Analysis of reaction rates is important for several applications, such as in chemical engineering or in chemical equilibrium study. Rates of reaction depends basically on:

- **Reactant concentrations,** which usually make the reaction happen at a faster rate if raised through increased collisions per Surface area available for contact between the reactants, in particular solid ones in heterogeneous systems. Larger surface area leads to higher reaction rates.
- **Pressure,** by increasing the pressure, you decrease the volume between molecules. This will increase the frequency of collisions of molecules.
- **Activation energy,** which is defined as the amount of energy required to make the reaction start and carry on spontaneously. Higher activation energy implies that the reactants need more energy to start than a reaction with a lower activation energy.
- **Temperature,** which hastens reactions if raised, since higher temperature increases the energy of the molecules, creating more collisions per unit time.
- The presence or absence of a *catalyst*. Catalysts are substances which change the pathway (mechanism) of a reaction which in turn increases the speed of a reaction by lowering the activation energy needed for the reaction to take place. A catalyst is not destroyed or changed during a reaction, so it can be used again.

  For some reactions, the presence of *electromagnetic radiation,* most notably ultraviolet, is needed to promote the breaking of bonds to start the reaction. This is particularly true for reactions involving radicals.

Reaction rates are related to the concentrations of substances involved in reactions, as quantified by the rate law of each reaction. Note that some reactions have rates that are *independent* of reactant concentrations. These are called zero order reactions.

Reactions and Energy

Chemical energy is part of all chemical reactions. Energy is needed to break chemical bonds in the starting substances. As new bonds form in the final substances, energy is released. By comparing the chemical energy of the original substances with the chemical energy of the final substances, you can decide if energy is released or absorbed in the overall reaction.

Exothermic Reactions

A chemical reaction in which energy is released is called an exothermic reaction. *Exo* means "go out" or "exit." *Thermic* means "heat" or "energy." Exothermic reactions can give off energy in several forms. If heat is released in an exothermic reaction, the nearby matter will become warmer. The nearby matter absorbs the heat released by the reaction. The reaction between gasoline and oxygen in a car's engine is an exothermic reaction

LIST OF ORGANIC REACTIONS

0-9 1, 3 – dipolar cycloaddition

A	Adkins catalyst
Abramovitch-Shapiro tryptamine synthesis	Adkins-Peterson reaction
	Akabori amino acid reaction
Acetoacetic ester condensation	Alder ene reaction
Achmatowicz reaction	Alder-Stein rules
Acyloin condensation	Aldol addition
Adams catalyst	Aldol condensation

Algar-Flynn-Oyamada reaction

Allan-Robinson reaction

Allylic rearrangement

Amadori rearrangement

Andrussov oxidation

Appel reaction

Arbuzov reaction, Arbusow reaction

Arens-van Dorp synthesis, Isler modification

Arndt-Eistert synthesis

Auwers synthesis

Azo coupling

B

Baeyer-Drewson indigo synthesis

Baeyer-Villiger oxidation

Baeyer-Villiger rearrangement

Bakeland process (Bakelite)

Baker-Venkataraman rearrangement

Baker-Venkataraman transformation

Bally-Scholl synthesis

Balz-Schiemann reaction

Bamberger rearrangement

Bamberger triazine synthesis

Bamford-Stevens reaction

Barbier-Wieland degradation

Bardhan-Senguph phenanthrene synthesis

Bartoli indole synthesis

Bartoli reaction

Barton reaction

Barton-McCombie reaction, Barton deoxygenation

Baudisch reaction

Bayer test

Baylis-Hillman reaction

Bechamp reaction

Beckmann fragmentation

Beckmann rearrangement

Bellus-Claisen rearrangement

Belousov-Zhabotinsky reaction

Benary reaction

Benzidine rearrangement

Benzilic acid rearrangement

Benzoin condensation

Bergman cyclization

Bergmann azlactone peptide synthesis

Bergmann degradation

Bergmann-Zervas carbobenzoxy method

Bernthsen acridine synthesis

Bestmann's reagent

Betti reaction

Biginelli pyrimidine synthesis

Biginelli reaction

Birch reduction

Bischler-Möhlau indole synthesis

Bischler-Napieralski reaction

Blaise ketone synthesis

Blaise reaction
Blanc reaction
Blanc chloromethylation
Bodroux reaction
Bodroux-Chichibabin aldehyde synthesis
Bogert-Cook synthesis
Bohn-Schmidt reaction
Boord olefin synthesis
Borodin reaction
Borsche-Drechsel cyclization
Bosch-Meiser urea process
Bouveault aldehyde synthesis
Bouveault-Blanc reduction
Boyland-Sims oxidation
Boyer Reaction
Bredt's rule
Brown hydroboration
Bucherer carbazole synthesis
Bucherer reaction
Bucherer-Bergs reaction
Buchner ring enlargement
Buchner-Curtius-Schlotterbeck reaction
Buchwald-Hartwig amination
Bunnett reaction

C

Cadiot-Chodkiewicz coupling
Camps quinoline synthesis
Cannizzaro reaction
Carroll reaction
Catalytic reforming
CBS reduction
Chan-Lam coupling
Chapman rearrangement
Chichibabin pyridine synthesis
Chichibabin reaction
Chugaev elimination
Ciamician-Dennstedt rearrangement
Claisen condensation
Claisen rearrangement
Claisen-Schmidt condensation
Clemmensen reduction
Collins-Reagent
Combes quinoline synthesis
Conia reaction
Conrad-Limpach synthesis
Corey-Gilman-Ganem oxidation
Cook-Heilbron thiazole synthesis
Cope elimination
Cope rearrangement
Corey reagent
Corey-Bakshi-Shibata reduction
Corey-Fuchs reaction
Corey-Kim oxidation
Corey-Posner, Whitesides-House reaction
Corey-Winter olefin synthesis
Corey-Winter reaction
Coupling reaction

Craig method

Cram's rule of asymmetric induction

Creighton process

Criegee reaction

Criegee rearrangement

Cross metathesis

Crum Brown-Gibson rule

Curtius degradation

Curtius rearrangement, Curtius reaction

D

Dakin reaction

Dakin-West reaction

Danheiser Annulation

Darapsky degradation

Darzens condensation, Darzens-Claisen reaction, Glycidic ester condensation

Darzens synthesis of unsaturated ketones

Darzens tetralin synthesis

Delepine reaction

Demjanov rearrangement

Demjanow desamination

Dess-Martin oxidation

Diazotisation

DIBAL-H selective reduction

Dieckmann condensation

Dieckmann reaction

Diels-Alder reaction

Diels Reese reaction

Dienol benzene rearrangement

Dienone phenol rearrangement

Dimroth rearrangement

Di-pi-methane rearrangement

Directed ortho metalation

Doebner modification

Doebner reaction

Doebner-Miller reaction, Beyer method for quinolines

Doering-LaFlamme carbon chain extension

Dötz reaction

Dowd-Beckwith ring expansion reaction

Duff reaction

Dutt-Wormall reaction

E

E1cB elimination reaction

Eder reaction

Edman degradation

Eglinton reaction

Ehrlich-Sachs reaction

Einhorn variant

Einhorn-Brunner reaction

Elbs persulfate oxidation

Elbs reaction

Elimination reaction

Emde degradation

Emmert reaction

Ene reaction

Epoxidation

Erlenmeyer synthesis, Azlactone synthesis

Erlenmeyer-Plöchl azlactone and amino acid synthesis

Eschenmoser fragmentation

Eschweiler-Clarke reaction

Ester pyrolysis

Étard reaction

Evans aldol

F

Favorskii reaction

Favorskii rearrangement

Favorskii-Babayan synthesis

Feist-Benary synthesis

Fenton reaction

Ferrario reaction

Ferrier rearrangement

Finkelstein reaction

Fischer indole synthesis

Fischer oxazole synthesis

Fischer peptide synthesis

Fischer phenylhydrazine and oxazone reaction

Fischer glycosidation

Fischer-Hepp rearrangement

Fischer-Speier esterification

Fischer Tropsch synthesis

Fleming-Tamao oxidation

Flood reaction

Forster reaction

Forster-Decker method

Franchimont reaction

Frankland synthesis

Frankland-Duppa reaction

Freund reaction

Friedel-Crafts Acylation

Friedel-Crafts Alkylation

Friedländer synthesis

Fries rearrangement

Fritsch-Buttenberg-Wiechell rearrangement

Fujimoto-Belleau reaction

Fukuyama coupling

Fukuyama indole synthesis

G

Gabriel ethylenimine method

Gabriel synthesis

Gabriel-Colman rearrangement, Gabriel isoquinoline synthesis

Gallagher-Hollander degradation

Gassman indole synthesis

Gastaldi synthesis

Gattermann aldehyde synthesis

Gattermann Koch reaction

Gattermann reaction

Gewald reaction

Gibbs phthalic anhydride process

Gilman reagent

Glaser coupling

Glycol cleavage

Gogte synthesis

Gomberg-Bachmann reaction

Gomberg-Bachmann-Hey reaction

Gomberg-Free radical reaction

Gould-Jacobs reaction

Graebe-Ullmann synthesis

Grignard degradation

Grignard reaction

Grob fragmentation

Grubbs' catalyst in Olefin metathesis

Grundmann aldehyde synthesis

Gryszkiewicz-Trochimowski and McCombie method

Guareschi-Thorpe condensation

Guerbet reaction

Gutknecht pyrazine synthesis

H

Haller-Bauer reaction

Haloform reaction

Hammett equation

Hammick reaction

Hammond-Principle or Hammond postulate

Hantzsch pyrrole synthesis

Hantzsch dihydropyridine synthesis, Hantzsch pyridine synthesis

Hantzsch Pyridine synthesis, Gattermann-Skita synthesis, Guareschi-Thorpe condensation, Knoevenagel-Fries modification

Hantzsch-Collidin-synthesis

Harber-Weiss reaction

Harries Ozonide reaction

Haworth Methylation

Haworth Phenanthrene synthesis

Haworth-reaction

Hay coupling

Hayashi rearrangement

Heck reaction

Helferich method

Hell-Volhard-Zelinsky halogenation

Hemetsberger indole synthesis

Hemetsberger-Knittel synthesis

Henkel reaction, Raecke process, Henkel process

Henry reaction, Kamlet reaction

Herz reaction, Herz compounds

Herzig-Meyer alkimide group determination

Heumann indigo synthesis

Hinsberg indole synthesis

Hinsberg oxindole synthesis

Hinsberg reaction

Hinsberg separation

Hinsberg sulfone synthesis

Hoch-Campbell ethylenimine synthesis

Hofmann degradation, Exhaustive methylation

Hofmann Elimination

Hofmann Isonitrile synthesis,

Carbylamine reaction
Hofmann produkt
Hofmann rearrangement
Hofmann-Löffler reaction, Löffler-Freytag reaction, Hofmann-Löffler-Freytag reaction
Hofmann-Martius rearrangement
Hofmann's Rule
Hofmann-Sand reaction
Homo rearrangement of steroids
Hooker reaction
Horner-Wadsworth-Emmons reaction
Hösch reaction
Hosomi-Sakurai reaction
Houben-Fischer synthesis
Hunsdiecker reaction
Hydroboration
Hydrohalogenation

I

Ing-Manske procedure
Ipso substitution
Ivanov reagent, Ivanov reaction

J

Jacobsen rearrangement
Janovsky reaction
Japp-Klingemann reaction
Japp-Maitland condensation
Johnson-Claisen rearrangement
Jones oxidation
Jordan-Ullmann-Goldberg synthesis
Julia olefination
Julia-Lythgoe olefination

K

Kabachnik-Fields reaction
Kendall-Mattox reaction
Kiliani-Fischer synthesis
Kindler reaction
Kishner cyclopropane synthesis
Knoevenagel condensation
Knoop-Oesterlin amino acid synthesis
Knorr pyrazole synthesis
Knorr pyrrole synthesis
Knorr quinoline synthesis
Koch-Haaf reaction
Kochi reaction
Koenigs-Knorr reaction
Kolbe electrolysis
Kolbe-Schmitt reaction
Kondakov rule
Kontanecki acylation
Kornblum oxidation
Kornblum–DeLaMare rearrangement
Kowalski ester homologation
Krafft degradation
Krapcho decarboxylation
Kröhnke aldehyde synthesis

Kröhnke oxidation

Kröhnke pyridine synthesis

Kucherov reaction

Kuhn-Winterstein reaction

Kulinkovich reaction

Kumada coupling

L

Larock indole synthesis

Lebedev process

Lehmstedt-Tanasescu reaction

Leimgruber-Batcho indole synthesis

Letts nitrile synthesis

Leuckart reaction

Leuckart thiophenol reaction

Leuckart-Wallach reaction

Leuckert amide synthesis

Levinstein process

Ley Oxidation

Lieben iodoform reaction, Haloform reaction

Liebeskind-Srogl coupling

Lindlar catalyst

Lobry-de Bruyn-van Ekenstein transformation

Lossen rearrangement

Luche reduction

M

Madelung synthesis

Malaprade reaction, Periodic acid oxidation

Malonic ester synthesis

Mannich reaction

Markovnikov's rule, Markownikoff rule, Markownikow rule

Martinet dioxindole synthesis

McDougall monoprotection

McFadyen-Stevens reaction

McMurry reaction

Meerwein arylation

Meerwein-Ponndorf-Verley reduction

Meisenheimer rearrangement

Meissenheimer complex

Menshutkin reaction

Mentzer pyrone synthesis

Metal-ion-catalyzed s-bond rearrangement

Mesylation

Merckwald asymmetric synthesis

Meyer and Hartmann reaction

Meyer reaction

Meyer synthesis

Meyer-Schuster rearrangement

Michael addition

Michael addition, Michael system

Michael condensation

Michaelis-Arbuzov reaction

Miescher degradation

Mignonac reaction

Milas hydroxylation of olefins

Mitsunobu reaction
Mukaiyama aldol addition
Mukaiyama reaction
Myers' asymmetric alkylation

N

Nametkin rearrangement
Nazarov cyclization reaction
Neber rearrangement
Nef reaction
Negishi coupling
Negishi-Zipper reaction
Nenitzescu indole synthesis
Nenitzescu reductive acylation
Nicholas reaction
Niementowski quinazoline synthesis
Niementowski quinoline synthesis
Nierenstein reaction
Nitroaldol reaction
Normant reagents
Noyori asymmetric hydrogenation
Nozaki-Hiyama-Kishi Nickel/ Chromium Coupling reaction
Nucleophilic acyl substitution

O

Ohira-Bestmann reaction
Olefin metathesis
Oppenauer oxidation
Ostromyslenskii reaction,
Ostromisslenskii reaction
Oxidative decarboxylation
Oxo synthesis
Oxy-Cope rearrangement
Oxymercuration
Ozonolysis

P

Paal-Knorr pyrrole synthesis
Paal-Knorr synthesis
Paneth technique
Paolini reaction
Passerini reaction
Paterno-Büchi reaction
Pauson-Khand reaction
Pechmann condensation
Pechmann pyrazole synthesis
Pellizzari reaction
Pelouze synthesis
Perkin alicyclic synthesis
Perkin reaction
Perkin rearrangement
Perkow reaction
Petasis reaction
Petasis reagent
Peterson olefination
Peterson reaction
Petrenko-Kritschenko piperidone synthesis
Pfan-Plattner azulene synthesis
Pfitzinger reaction

Pfitzner-Moffatt oxidation

Pictet-Gams isoquinoline synthesis

Pictet-Hubert reaction

Pictet-Spengler tetrahydroisoquinoline synthesis

Pictet-Spengler reaction

Piloty alloxazine synthesis

Piloty-Robinson pyrrole synthesis

Pinacol coupling reaction

Pinacol rearrangement

Pinner amidine synthesis

Pinner method for ortho esters

Pinner reaction

Pinner triazine synthesis

Piria reaction

Pitzer strain

Polonovski reaction

Pomeranz-Fritsch reaction

Ponzio reaction

Prelog strain

Prevost reaction

Prileschajew reaction

Prilezhaev reaction

Prins reaction

Prinzbach synthesis

Pschorr reaction

Pummerer rearrangement

Purdie methylation, Irvine-Purdie methylation

Q

Quelet reaction

R

Ramberg-Backlund reaction

Raney-Nickel

Rap-Stoermer condensation

Raschig phenol process

Rauhut-Currier reaction

Reed reaction

Reformatskii reaction

Reilly-Hickinbottom rearrangement

Reimer-Tiemann reaction

Reissert indole synthesis

Reissert reaction, Reissert compound

Reppe synthesis

Retropinacol rearrangement

Reverdin reaction

Riehm quinoline synthesis

Riemschneider thiocarbamate synthesis

Riley oxidations

Ring closing metathesis

Ring opening metathesis

Ritter reaction

Robinson annulation

Robinson-Gabriel synthesis

Robinson Schopf reaction

Rosenmund reaction

Rosenmund reduction
Rosenmund-von Braun synthesis
Rothemund reaction
Rowe rearrangement
Rupe reaction
Rubottom oxidation
Ruff-Fenton degradation
Ruzicka large ring synthesis

S

Sakurai reaction
Salol reaction
Sandheimer
Sandmeyer diphenylurea isatin synthesis
Sandmeyer isonitrosoacetanilide isatin synthesis
Sandmeyer reaction
Sanger reagent
Sarett oxidation
Saytzeff rule, Saytzeff's Rule
Schiemann reaction
Schlenk equilibrium
Schlosser modification
Schlosser variant
Schlosser-Lochmann reaction
Schmidlin ketene synthesis
Schmidt degradation
Schmidt reaction
Scholl reaction
Schorigin Shorygin reaction, Shorygin reaction, Wanklyn reaction
Schotten-Baumann reaction
Screttas-Yus reaction
Semidine rearrangement
Semmler-Wolff reaction
Serini reaction
Seyferth-Gilbert homologation
Shapiro reaction
Sharpless asymmetric dihydroxylation
Sharpless epoxidation
Sharpless oxyamination or aminohydroxylation
Simmons-Smith reaction
Simonini reaction
Simonis chromone cyclization
Skraup chinolin synthesis
Skraup reaction
Smiles rearrangement
SNAr nucleophilic aromatic substitution
SN1
SN2
SNi
Sommelet reaction
Sonn-Müller method
Sonogashira coupling
Sørensen formol titration
Staedel-Rugheimer pyrazine synthesis
Staudinger reaction
Stephen aldehyde synthesis

Stetter reaction
Stevens rearrangement
Stieglitz rearrangement
Stille coupling
Stobbe condensation
Stollé synthesis
Stork acylation
Stork enamine alkylation
Strecker amino acid synthesis
Strecker degradation
Strecker sulfite alkylation
Strecker synthesis
Stuffer disulfone hydrolysis rule
Suzuki coupling
Swain equation
Swarts reaction
Swern oxidation

T

Tamao oxidation
Tafel rearrangement
Takai olefination
Tebbe olefination
Ter Meer reaction
Thermite reactions
Thiele reaction
Thorpe reaction
Tiemann rearrangement
Tiffeneau ring enlargement reaction
Tiffeneau-Demjanow rearrangement
Tischtschenko reaction
Tishchenko reaction, Tischischenko-Claisen reaction
Tollens reagent
Trapp mixture
Traube purine synthesis
Truce-Smiles rearrangement
Tscherniac-Einhorn reaction
Tschitschibabin reaction
Tschugajeff reaction
Twitchell process
Tyrer sulfonation process

U

Ugi reaction
Ullmann reaction
Upjohn dihydroxylation
Urech cyanohydrin method
Urech hydantoin synthesis

V

Van Slyke determination
Varrentrapp reaction
Vilsmeier reaction
Vilsmeier-Haack reaction
Voight amination
Volhard-Erdmann cyclization
von Braun amide degradation
von Braun reaction
von Richter cinnoline synthesis
von Richter reaction

W	Wohl-Aue reaction
Wacker-Tsuji oxidation	Wohler synthesis
Wagner-Jauregg reaction	Wohl-Ziegler reaction
Wagner-Meerwein rearrangement	Wolffenstein-Böters reaction
Walden inversion	Wolff rearrangement
Wallach rearrangement	Wolff-Kishner reduction
Weerman degradation	Woodward cis-hydroxylation
Weinreb ketone synthesis	Woodward-Hoffmann rule
Wenker ring closure	Wulff-Dötz reaction
Wenker synthesis	Wurtz coupling, Wurtz reaction
Wessely-Moser rearrangement	Wurtz-Fittig reaction
Westphalen-Lettré rearrangement	**Y**
Wharton reaction	Yamaguchi esterification
Whiting reaction	**Z**
Wichterle reaction	Zeisel determination
Widman-Stoermer synthesis	Zerevitinov determination, Zerewitinoff determination
Wilkinson catalyst	Ziegler condensation
Willgerodt rearrangement	Ziegler method
Willgerodt-Kindler reaction	Zimmermann reaction
Williamson ether synthesis	Zincke disulfide cleavage
Winstein reaction	Zinke nitration
Wittig reaction	Zincke reaction
Wittig rearrangement	Zincke-Suhl reaction
Wittig-Horner reaction	
Wohl degradation	

6

Organic Reaction

Organic reactions are chemical reactions involving organic compounds The basic organic chemistry reaction types are addition reactions, elimination reactions, substitution reactions, pericyclic reactions, rearrangement reactions and redox reactions. In organic synthesis, organic reactions are used in the construction of new organic molecules. The production of many man-made chemicals such as drugs, plastics, food additives, fabrics depend on organic reactions.

The oldest organic reactions are combustion of organic fuels and saponification of fats to make soap. Modern organic chemistry starts with the Wöhler synthesis in 1828. In the history of the Nobel Prize in Chemistry awards have been given for the invention of specific organic reactions such as the Grignard reaction in 1912, the Diels-Alder reaction in 1950, the Wittig reaction in 1979 and olefin metathesis in 2005.

Classifications

Organic chemistry has a strong tradition of naming a specific reaction to its inventor or inventors and a long list of so-called named reactions exists, conservatively estimated at 1000. A very old named reaction is the Claisen rearrangement (1912) and a recent named reaction is the Bingel reaction (1993). When the named reaction is difficult to pronounce or very long as in the Corey-House-Posner-Whitesides reaction it helps to use the abbreviation as in the CBS reduction. The number of reactions hinting at the actual process taking place is much smaller, for example the ene reaction or aldol reaction.

Another approach to organic reactions is by type of organic reagent, many of them inorganic, required in a specific transformation. The major types are oxidizing agents such as osmium tetroxide, reducing agents such as Lithium aluminium hydride, bases such as lithium diisopropylamide and acids such as sulfuric acid.

Fundamentals

Factors governing organic reactions are essentially the same as that of any chemical reaction. Factors specific to organic reactions are those that determine the stability of reactants and products such as conjugation, hyperconjugation and aromaticity and the presence and stability of reactive intermediates such as free radicals, carbocations and carbanions.

An organic compound may consist of many isomers. Selectivity in terms of regioselectivity, diastereoselectivity and enantioselectivity is therefore an important criterion for many organic reactions. The stereochemistry of pericyclic reactions is governed by the Woodward-Hoffmann rules and that of many elimination reactions by the Zaitsev's rule.

Organic reactions are important in the production of pharmaceuticals. In a 2006 review it was estimated that 20% of chemical conversions involved alkylations on nitrogen and oxygen atoms, another 20% involved placement and removal of protective groups, 11% involved formation of new carbon-carbon bond and 10% involved functional group interconversions.

Organic Reactions by Mechanism

There is no limit to the number of possible organic reactions and mechanisms. However, certain general patterns are observed that can be used to describe many common or useful reactions. Each reaction has a stepwise reaction mechanism that explains how it happens, although this detailed description of steps is not always clear from a list of reactants alone. Organic reactions can be organized into several basic types. Some reactions fit into more than one category. For example, some substitution reactions follow an addition-elimination pathway. This overview isn't intended to include every single organic reaction. Rather, it is intended to cover the basic reactions.

Addition reactions include such reactions as halogenation, hydrohalogenation and hydration. The main addition reactions are:

- electrophilic addition or EA
- nucleophilic addition or NA
- radical addition or RA

Elimination reactions include processes such as dehydration and are found to follow an E_1, E_2 or E_1cB reaction mechanism.

Substitution reactions are divided into:

- nucleophilic aliphatic substitution with S_N1, S_N2 and S_Ni reaction mechanisms
- nucleophilic aromatic substitution or NAS
- nucleophilic acyl substitution
- electrophilic substitution or ES
- electrophilic aromatic substitution or EAS
- radical substitution or RS

Organic redox reactions are redox reactions specific to organic compounds and are very common.

Rearrangement reactions are divided into:

- 1, 2-rearrangements.
- pericyclic reactions.
- metathesis

In *Condensation reactions* a small molecule, usually water, is split off when two reactants combine in a chemical reaction. The opposite reaction, when water is consumed in a reaction, is called hydrolysis. Many Polymerization reactions are derived from organic reactions. They are divided into addition polymerizations and step-growth polymerizations.

Organic Reactions by Functional Groups

Organic reactions can be categorized based on the type of functional group involved in the reaction as a reactant and the functional group that is formed as a result of this reaction. For example in the Fries rearrangement the reactant is an ester and the reaction product an alcohol.

Other Organic Reaction Classification

In heterocyclic chemistry, organic reactions are classified by the type of heterocycle formed with respect to ring-size and type of heteroatom.

Organic reactions can also be classified by the type of bond to carbon with respect to the element involved. More reactions are found in organosilicon chemistry, organosulfur chemistry, organophosphorus chemistry and organofluorine chemistry. With the introduction of carbon-metal bonds the field crosses over to organometallic chemistry.

Addition Reaction

An addition reaction, in organic chemistry, is in its simplest terms an organic reaction where two or more molecules combine to form a larger one.

There are two main types of polar addition reactions:

- Electrophilic addition
- Nucleophilic addition
- Other non-polar addition reactions exists as well called free radical addition

Addition reactions are limited to chemical compounds that have multiply-bonded atoms:

- Molecules with carbon-carbon double bonds (alkenes) or triple bonds (alkynes)
- Molecules with carbon - hetero double bonds like C=O or C=N

An addition reaction is the opposite of an elimination reaction. For instance the hydration reaction of an alkene and the dehydration of an alcohol are addition-elimination pairs.

Addition-elimination Reaction

In the related Addition-elimination reaction an addition reaction is followed by an elimination reaction. In the majority of reactions it involves addition of nucleophiles to carbonyl compounds in what is called nucleophilic acyl substitution.

Other addition-elimination reactions are:

- reaction of an aliphatic amine to an imine and an aromatic amine to a Schiff base in alkylimino-de-oxo-bisubstitution
- Hydrolysis of nitriles to carboxylic acids

Electrophilic Addition

In organic chemistry, an electrophilic addition reaction is an addition reaction where, in a chemical compound, a pi bond is removed by the creation of two new covalent bonds. The substrate of an electrophilic addition reaction must have a double bond or triple bond.

The driving force for this reaction is the formation of an electrophile X^+ that forms a covalent bond with an electron-rich unsaturated C=C bond. The positive charge on X is transferred to the carbon-carbon bond.

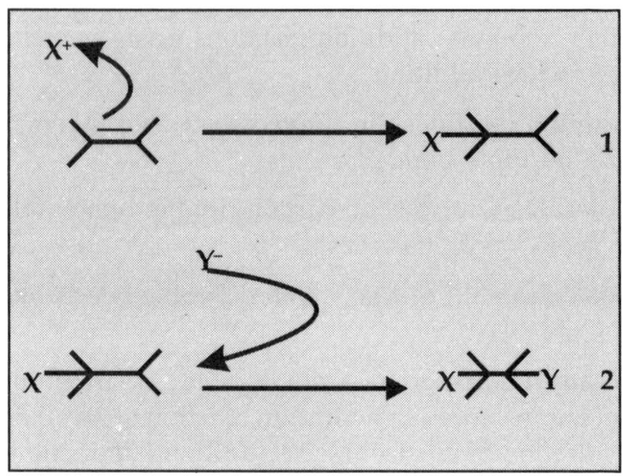

Fig. 6.1

In step 2 of an electrophilic addition, the positively charged intermediate combines with (Y) that is electron-rich and usually an anion to form the second covalent bond.

Step 2 is the same nucleophilic attack process found in an S_N1 reaction. The exact nature of the electrophile and the nature of the positively charged intermediate are not always clear and depend on reactants and reaction conditions.

In all asymmetric addition reactions to carbon, regioselectivity is important and often determined by Markovnikov's rule. Organoborane compounds give anti-Markovnikov additions. Electrophilic attack to an aromatic system results in electrophilic aromatic substitution rather than an addition reaction.

Typical Electrophilic Additions

Typical electrophilic additions to alkenes with reagents are:

- dihalo addition reactions: X_2

- Hydrohalogenations: HX
- Hydration reactions: H_2O
- Hydrogenations H_2
- Oxymercuration reactions: mercuric acetate, water
- Hydroboration-oxidation reactions : diborane the Prins reaction: formaldehyde, water.

Elimination Reaction

An elimination reaction is a type of organic reaction in which two substituents are removed from a molecule in either a one or two-step mechanism Either the unsaturation of the molecule increases (as in most organic elimination reactions) or the valence of an atom in the molecule decreases by two, a process known as reductive elimination.

An important class of elimination reactions are those involving alkyl halides, or alkanes in general, with good leaving groups, reacting with a Lewis base to form an alkene in the reverse of an addition reaction. When the substrate is asymmetric, regioselectivity is determined by Zaitsev's rule. The one and two-step mechanisms are named and known as E2 reaction and E1 reaction, respectively.

E2 Mechanism

In the 1920s, Sir Christopher Ingold proposed a model to explain a peculiar type of chemical reaction: the E2 mechanism. E2 stands for bimolecular elimination and has the following specificities.

- It is a one-step process of elimination with a single transition state.
- Typical of secondary or tertiary substituted alkyl halides. It is also observable with primary alkyl halides if a hindered base is used.
- The reaction rate, influenced by both the alkyl halide and the base, is second order.

- Because E2 mechanism results in formation of a pi bond, the two leaving groups (often a hydrogen and a halogen) need to be coplanar. An antiperiplanar transition state has staggered conformation with lower energy and a synperiplanar transition state is in eclipsed conformation with higher energy. The reaction mechanism involving staggered conformation is more favourable for E2 reactions.
- Reaction often present with strong base.
- In order for the pi bond to be created, the hybridization of carbons need to be lowered from sp^3 to sp^2.
- The C-H bond is weakened in the rate determining step and therefore the deuterium isotope effect is larger than 1.
- This reaction type has similarities with the SN2 reaction mechanism.

  The reaction fundamental elements are:
- Breaking of the *carbon-hydrogen* and *carbon-halogen* bonds in one step.

  Formation of a C = C *Pi bond*

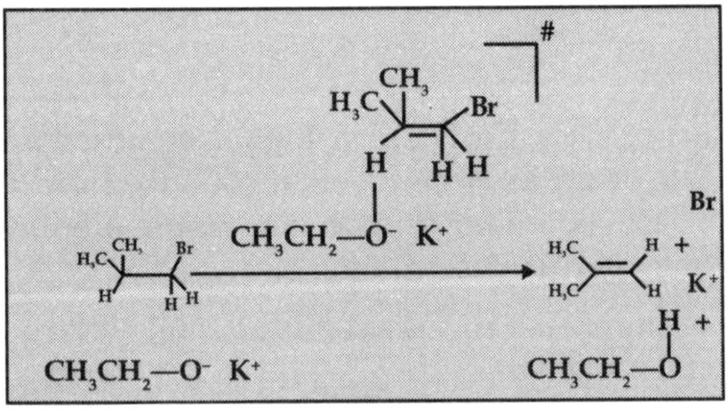

Fig. 6.2

An example of this type of reaction in *scheme 1* is the reaction of isobutylbromide with potassium ethoxide in ethanol. The reaction products are isobutylene, ethanol and potassium bromide.

E1 Mechanism

E1 is a model to explain a particular type of chemical elimination reaction. E1 stands for unimolecular elimination and has the following specificities.

- It is a two-step process of elimination *ionization and deprotonation*.
- Ionization, Carbon-halogen breaks to give a carbocation intermediate.
- Deprotonation of the carbocation.
- Typical of tertiary and some secondary substituted alkyl halides.
- The reaction rate is influenced only by the concentration of the alkyl halide because carbocation formation is the slowest, rate-determining step. Therefore first order kinetics apply.
- Reaction mostly occurs in complete absence of base or presence of only a weak base.
- E1 reactions are in competition with SN1 reactions because they share a common carbocationic intermediate.
- A deuterium isotope effect is absent.
- No antiperiplanar requirement. An example is the pyrolysis of a certain sulfonate ester of menthol:

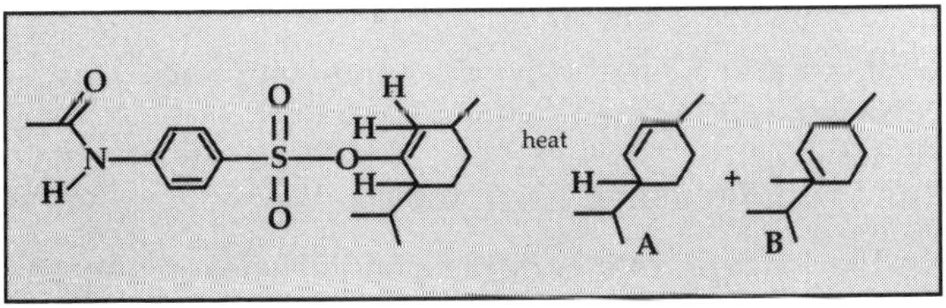

Fig. 6.3

Only reaction product A results from antiperiplanar elimination, the presence of product B is an indicator for a E1 mechanism

- Accompanied by carbocationic rearrangement reactions

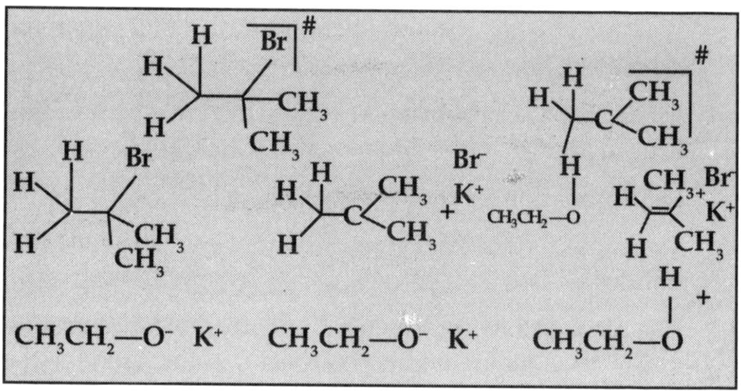

Fig. 6.4

An example in *scheme 2* is the reaction of tert-butylbromide with potassium ethoxide in ethanol.

E1 eliminations happen with highly substituted alkyl halides due to 2 main reasons.

- Highly substituted alkyl halides are bulky, limiting the room for the E2 one-step mechanism; therefore, the two-step E1 mechanism is favored.
- Highly substituted carbocations are more stable than methyl or primary substituted. Such stability gives time for the two-step E1 mechanism to occur.

If S_N1 and E1 pathways are competing, the E1 pathway can be favored by increasing the heat.

E2 and E1 Elimination Final Notes

The reaction rate is influenced by halogen's reactivity; iodide and bromide being favored. Fluoride is not a good leaving group. There is a certain level of competition between elimination reaction and nucleophilic substitution. More precisely, there are competitions between E2 and S_N2 and also between E1 and SN1. Substitution generally predominates and elimination occurs only during precise circumstances. Generally, elimination is favored over substitution when

- steric hindrance increases
- basicity increases
- temperature increases
- the steric bulk of the base increases for example Potassium tert-butoxide
- the nucleophile is poor

In one study the kinetic isotope effect (KIE) was determined for the gas phase reaction of several alkyl halides with the chlorate ion. In accordance with a E2 elimination the reaction with t-butyl chloride results in a KIE of 2.3. The methyl chloride reaction (only S_N2 possible) on the other hand has a KIE of 0.85 consistent with a S_N2 reaction because in this reaction type the C-H bonds tighten in the transition state. The KIE's for the ethyl (0.99) and isopropyl (1.72) analogues suggest competition between the two reaction modes.

Specific Elimination Reactions

The E1cB elimination reaction is a special type of elimination reaction involving carbanions. In an addition-elimination reaction elimination takes place after an initial addition reaction and in the Ei mechanism both substituents leave simultaneously in a syn addition.

In each of these elimination reactions the reactants have specific leaving groups:

- the dehydration reaction is one where the leaving group is water.
- the Bamford-Stevens reaction with a tosyl hydrazone leaving group assisted by alkoxide.

- the Cope reaction with an amine oxide leaving group.
- the Hofmann elimination with quaternary amine leaving group.
- the Chugaev reaction with a methyl xanthate leaving group.
- the Grieco elimination with a selenoxide leaving group.
- the Shapiro reaction with a tosyl hydrazone leaving group assisted by alkyllithium.
- Hydrazone iodination with a hydrazone leaving group assisted by iodine
- A Grob fragmentation with degree of unsaturation increasing in one of the leaving groups.
- the Kornblum–DeLaMare rearrangement (elimination over a (H)C-O(OR) bond) with an alcohol leaving group forming a ketone
- the Takai olefination with two bulky chromium groups.

Substitution Reaction

In a substitution reaction, a functional group in a particular chemical compound is replaced by another group In organic chemistry, the electrophilic and nucleophilic substitution reactions are of prime importance. Organic substitution reactions are classified in several main organic reaction types depending on whether the reagent that brings about the substitution is considered an electrophile or a nucleophile, whether a reactive intermediate involved in the reaction is a carbocation, a carbanion or a free radical or whether the substrate is aliphatic or aromatic. Detailed understanding of a reaction type helps to predict the product outcome in a reaction. It also is helpful for optimizing a reaction with regard to variables such as temperature and choice of solvent.

A good example of a substitution reaction is the photochemical chlorination of methane forming methyl chloride:

$$CH_4 + Cl\text{-}Cl \xrightarrow{h\nu} CH_3Cl + H\text{-}Cl$$

Chlorination of methane by chlorine

Nucleophilic Substitutions

These kind of substitution reactions happen when the reagent is a nucleophile, which means, an atom or molecule with free electrons.

- A nucleophile reacts with an aliphatic substrate in a nucleophilic aliphatic substitution reaction.
- When the substrate is an aromatic compound the reaction type is nucleophilic aromatic substitution.
- Carboxylic acid derivatives react with nucleophiles in nucleophilic acyl substitution. This kind of reaction can be useful in preparing compounds.

The Nucleophilic substitutions can be produced by two different mechanisms:

- Monomolecular nucleophilic substitution (S_N1): In this case the reaction proceeds in stages, the compounds first dissociate in their ions and then this ions react between them. It's produced by carbocations.
- Bimolecular nucleophilic substitution (S_N2): In this case the reaction proceeds in only one stage. The attack of the reagent and the expulsion of the leaving group happen simultaneously.

Organic Redox Reaction

Organic reductions or organic oxidations or organic redox reactions are redox reactions that take place with organic compounds. In organic chemistry oxidations and reductions are different from ordinary redox reactions because many reactions carry the name but do not actually involve electron transfer in the electrochemical sense of the word.

Following the rules for determining the oxidation number for an individual carbon atom leads to:

- oxidation number -4 for alkanes;
- oxidation number -2 for alkenes, alcohols, alkyl halides, amines;
- oxidation number 0 for alkynes, ketones, aldehydes, geminal diols;
- oxidation number +2 for carboxylic acids, amides, chloroform and;
- oxidation number +4 for carbon dioxide, tetrachloromethane.

Methane is oxidized to carbon dioxide because the oxidation number changes from -4 to +4. Classical reductions include alkene reduction to alkanes and classical oxidations include oxidation of alcohols to aldehydes with manganese dioxide. In oxidations electrons are removed and the electron density of a molecule is reduced. In reductions electron density increases when electrons are added to the molecule. This terminology is always centered around the organic compound. So a ketone is always reduced by Lithium aluminium hydride but it is bad form to have lithium aluminium hydride oxidized by a ketone. Many oxidations involve removal of protons from the organic molecule and the reverse reduction adds protons to an organic molecule.

Many reactions classified as reductions also appear in other classes. For instance conversion of the ketone to an alcohol by Lithium aluminium hydride can be considered a reduction but the hydride is also a good nucleophile in nucleophilic substitution. Many redox reactions in organic chemistry have coupling reaction reaction mechanism involving free radical intermediates. True organic redox chemistry can be found in electrochemical organic synthesis or electrosynthesis. Examples of organic reactions that can take place in an electrochemical cell are the Kolbe electrolysis.

In disproportionation reactions the reactant is both oxidised and reduced in the same chemical reaction forming 2 separate compounds.

Asymmetric catalytic reductions and asymmetric catalytic oxidations are important in asymmetric synthesis.

Organic Reductions

Several reaction mechanisms exist for organic reductions:

- Direct electron transfer in one-electron reduction with the Birch reduction as example.
- Hydride transfer in reductions with for example lithium aluminium hydride or a hydride shift as in the Meerwein-Ponndorf-Verley reduction.
- Hydrogen reductions with a catalyst such as the Lindlar catalyst or the Adkins catalyst or in specific reductions such as the Rosenmund reduction.
- Disproportionation reaction such as the Cannizzaro reaction.

Reductions that do not fit in any reduction reaction mechanism and in which just the change in oxidation state is reflected include the Wolff-Kishner reaction.

Organic Oxidations

Several reaction mechanisms exist for organic oxidations:

- Single electron transfer
- Oxidations through ester intermediates with chromic acid or manganese dioxide

- Hydrogen atom transfer as in Free radical halogenation
- Oxidation with oxygen (combustion)
- Oxidation involving ozone in ozonolysis and peroxides
- Oxidations involving an elimination reaction mechanism such as the Swern oxidation, the Kornblum oxidation and with reagents such as IBX acid and Dess-Martin periodinane.
- oxidation by nitroso radicals Fremy's salt or tempo

Rearrangement Reaction

A rearrangement reaction is a broad class of organic reactions where the carbon skeleton of a molecule is rearranged to give a structural isomer of the original molecule. Often a substituent moves from one atom to another atom in the same molecule. In the example below the substituent R moves from carbon atom 1 to carbon atom 2:

```
C— C— C—   →   —C— C— C—
|                   |
R                   R
```

Intermolecular rearrangements also take place.

A rearrangement is not well represented by simple and discrete electron transfers (represented by curly arrows in organic chemistry texts). The actual mechanism of alkyl groups moving, as in Wagner-Meerwein rearrangement, probably involves transfer of the moving alkyl group fluidly along a bond, not ionic bond-breaking and forming. In pericyclic reactions, explanation by orbital interactions give a better picture than simple discrete electron transfers. It is, nevertheless, possible to draw the curly arrows for a sequence of discrete electron transfers that give the same result as a rearrangement reaction, although these are not necessarily realistic.

Some key rearrangement reactions:

- 1, 2-rearrangements
- pericyclic reactions
- olefin metathesis

7

Autocatalytic Reaction

Autocatalytic reactions are chemical reactions in which at least one of the products is also a reactant. The rate equations for autocatalytic reactions are fundamentally nonlinear. This nonlinearity can lead to the spontaneous generation of order. A dramatic example of this order is that which is found in living systems. This spontaneous order creation seems to contradict the Second Law of Thermodynamics. This contradiction is resolved when the disorder of both the system and its surroundings are taken into account.

Background

The Second Law of Thermodynamics states that the disorder (entropy) of a physical or chemical system and its surroundings must increase with time. In other words, systems left to themselves must become increasingly random. To say it yet another way, orderly energy of a system like uniform motion must degrade eventually to the random motion of particles in a heat bath.

This seems to run counter to experience. There are many instances in which physical systems spontaneously become emergent or orderly. For example, despite the destruction they do, hurricanes have a very orderly vortex motion when compared with the random motion of the air molecules in a closed room. Even more spectacular is the order created by chemical systems; the most dramatic being the order associated with life.

Our experience is consistent with the Second Law. The Second Law states that the total disorder of a system and its surrounding must increase with time. Order can be created in a system by an even greater decrease in order of the systems surroundings. In the hurricane example, hurricanes are formed from unequal heating within the atmosphere. The Earth's atmosphere is then far from thermal equilibrium. The order of the Earth's atmosphere increases, but at the expense of the order of the sun. The sun is becoming more disorderly as it ages and throws off light and material to the rest of the universe. The total disorder of the sun and the earth increases despite the fact that orderly hurricanes are generated on earth.

A similar example exists for living chemical systems. The sun provides energy to green plants. The green plants are food for other living chemical systems. The energy absorbed by plants and converted into chemical energy generates a system on earth that is orderly and far from chemical equilibrium. Here, the difference from chemical equilibrium is determined by an excess of reactants over the equilibrium amount. Once again, order on earth is generated at the expense of entropy increase of the sun. The total entropy of the earth and the rest of the universe increases, consistent with the Second Law.

Not all chemical reactions, however, generate order. The class of reactions most closely associated with order creation is the class of autocatalytic reactions. These are reactions in which one or more of the products are the same as one or more of the reactants. Simple autocatalytic reactions (clock reactions) are known to oscillate in time, thus creating temporal order. Other simple reactions can generate spatial separation of chemical species generating spatial order. More complex reactions are involved in metabolic pathways and metabolic networks in biological systems.

The transition to order as the distance from equilibrium increases is not usually continuous. Order typically appears abruptly. The threshold between the disorder of chemical equilibrium and order is known as a phase transition. The conditions for a phase transition can be determined with the mathematical machinery of non-equilibrium thermodynamics.

Chemical Reactions

A chemical reaction of two reactants and two products can be written as

$$\alpha A + \beta B \rightleftharpoons \sigma S + \tau T$$

where the Greek letters are stoichiometric coefficients and the capital Latin letters represent chemical species. The chemical reaction proceeds in both the forward and reverse direction. This equation is easily generalized to any number of reactants, products, and reactions.

Chemical Equilibrium

In chemical equilibrium the forward and reverse reaction rates are such that each chemical species is being created at the same rate it is being destroyed. In other words, the rate of the forward reaction is equal to the rate of the reverse reaction.

$$k_+ \{A\}^\alpha \{B\}^\beta = k_- \{S\}^\sigma \{T\}^\tau$$

Here, the curly brackets indicate the amount of the chemical species, in moles, and k_+ and k_- are rate constants.

Far from Equilibrium

Far from equilibrium, the forward and reverse reaction rates no longer balance and the concentration of reactants and products is no longer constant. For every forward reaction a molecules of A are destroyed. For every reverse reaction a molecules of A are created. The change in number of moles of A is then

$$\frac{d}{dt}\{A\} = -\alpha k_+\{A\}^\alpha\{B\}^\beta + \alpha k_-\{S\}^\sigma\{T\}^\tau$$

and similarly for the other reactant and products.

$$\frac{d}{dt}\{B\} = -\beta k_+\{A\}^\alpha\{B\}^\beta + \beta k_-\{S\}^\sigma\{T\}^\tau$$

$$\frac{d}{dt}\{S\} = -\sigma k_+\{A\}^\alpha\{B\}^\beta - \sigma k_-\{S\}^\sigma\{T\}^\tau$$

$$\frac{d}{dt}\{T\} = -\tau k_+\{A\}^\alpha\{B\}^\beta - \tau k_-\{S\}^\sigma\{T\}^\tau$$

This system of equations has a single stable fixed point when the forward rates and the reverse rates are equal. This means that the system evolves to the equilibrium state, and this is the only state to which it evolves.

Autocatalytic Reactions

Autocatalytic reactions are those in which at least one of the products is a reactant. Perhaps the simplest autocatalytic reaction can be written

$$A + B \rightleftharpoons 2B$$

with the rate equations

$$\frac{d}{dt}\{A\} = -k_+\{A\}\{B\} + k_-\{B\}^2$$

$$\frac{d}{dt}\{B\} = k_+\{A\}\{B\} - k_-\{B\}^2$$

This reaction is one in which a molecule of species A interacts with a molecule of species B. The A molecule is converted into a B molecule. The final product consists of the original B molecule plus the B molecule created in the reaction.

The key feature of these rate equations is that they are nonlinear; the second term on the right varies as the square of the concentration of B. This feature can lead to multiple fixed points of the system, much like a quadratic equation can have multiple roots. Multiple fixed points allow for multiple states of the system. A system existing in multiple macroscopic states is more orderly (has lower entropy) than a system in a single state.

CREATION OF ORDER

Temporal Order

Idealized Example: Lotka-Volterra Equation

Consider a coupled set of two autocatalytic reactions in which the concentration of one of the reactants A is much larger than its equilibrium value. In this case the forward reaction rate is so much larger than the reverse rates that we can neglect the reverse rates.

$$A + X \rightarrow 2X$$

$$X + Y \rightarrow 2Y$$

$$Y \rightarrow E$$

with the rate equations

$$\frac{d}{dt}\{X\} = k_1\{A\}\{X\} - k_2\{X\}\{Y\}$$

$$\frac{d}{dt}\{Y\} = k_2\{X\}\{Y\} - k_3\{Y\}$$

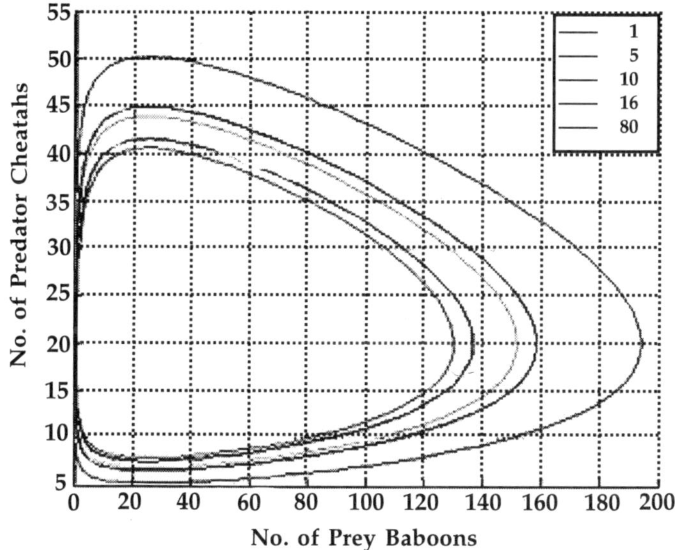

Fig 7.1

The Lotka-Volterra equation is isomorphic with the predator prey model and the two reaction autocatalytic model. In this example baboons and cheatahs are equivalent to two different chemical species in autocatalytic reactions

Here, we have neglected the depletion of the reactant A, since its concentration is so large. The rate constants for the three reactions are k_1, k_2, and k_3, respectively.

This system of rate equations is known as the Lotka-Volterra equation and is most closely associated with population dynamics in predator-prey relationships. This system of equations has an oscillatory behavior. The amplitude of the oscillations depends on the concentration of A. Oscillations of this type are a form of emergent temporal order that is not present in equilibrium.

Another idealized example: Brusselator

Another example of a system that demonstrates temporal order is the Brusselator (see Prigogine reference). It is characterized by the reactions.

$$A \rightarrow X$$

$$2X + Y \rightarrow 3X$$

$$B + X \rightarrow Y + D$$

$$X \rightarrow E$$

with the rate equations

$$\frac{d}{dt}\{Y\} = \{A\} + \{X\}^2\{Y\} - \{B\}\{X\} - \{X\}$$

$$\frac{d}{dt}\{Y\} = \{B\}\{X\} - \{X\}^2\{Y\}$$

where, for convenience, the rate constants have been set to 1.

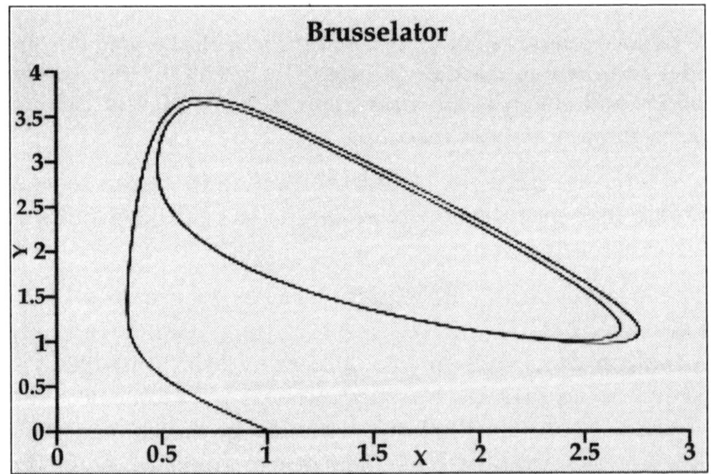

Fig 7.2

The Brusselator in the unstable regime. A=1. B=2.5. X(0)=1. Y(0)=0. The system approaches a limit cycle. For B<1+A the system is stable and approaches a fixed point

The Brusselator has a fixed point at

$$\{X\} = A$$

$$\{Y\} = \frac{B}{A}$$

The fixed point becomes unstable when

$$B > 1 + A^2$$

leading to an oscillation of the system. Unlike the Lotka-Volterra equation, the oscillations of the Brusselator do not depend on the amount of reactant present initially. Instead, after sufficient time, the oscillations approach a limit cycle.

Real Examples

Real examples of clock reactions are the Belousov-Zhabotinsky reaction (BZ reaction), the Briggs-Rauscher reaction, the Bray-Liebhafsky reaction and the iodine clock reaction. These are oscillatory reactions, and the concentration of products and reactants can be approximated in terms of damped oscillations.

The best-known reaction, the BZ reaction, can be created with a mixture of potassium bromate ($KBrO_3$), malonic acid ($CH_2(COOH)_2$), and manganese sulfate ($MnSO_4$) prepared in a heated solution of sulfuric acid (H_2SO_4).

Spatial Order

An idealized example of spatial spontaneous symmetry breaking is the case in which we have two boxes of material separated by a permeable membrane so that material can diffuse between the two boxes. It is assumed that identical Brusselators are in each box with nearly identical initial conditions. (see Prigogine reference)

$$\frac{d}{dt}\{X_1\} = \{A\} + \{X_1\}^2\{Y_1\} - \{B\}\{X_1\} - \{X_1\} + D_x\left(X_2 - X_1\right)$$

$$\frac{d}{dt}\{Y_1\} = \{B\} + \{X_1\} - \{X_1\}^2\{Y_1\} + D_y\left(Y_2 - Y_1\right)$$

$$\frac{d}{dt}\{X_2\} = \{A\} + \{X_2\}^2\{Y_2\} - \{B\}\{X_2\} - \{X_2\} + D_x\left(X_1 - X_2\right)$$

$$\frac{d}{dt}\{Y_2\} = \{B\} + \{X_2\} - \{X_2\}^2\{Y_2\} + D_y\left(Y_1 - Y_2\right)$$

Here, the numerical subscripts indicate which box the material is in. There are additional terms proportional to the diffusion coefficient D that account for the exchange of material between boxes.

If the system is initiated with the same conditions in each box, then a small fluctuation will lead to separation of materials between the two boxes. One box will have a predominance of X, and the other will have a predominance of Y.

Biological Example

It is known that an important metabolic cycle, glycolysis, displays temporal order Glycolysis consists of the degradation of one molecule of glucose and the overall production of two molecules of ATP. The process is therefore of great importance to the energetics of living cells. The global glycolysis reaction involves glucose, ADP, NAD, pyruvate, ATP, and NADH.

glucose + 2ADP + $2P_i$ + 2NAD $\rightarrow$ 2(pyruvate) + 2ATP + 2NADH

The details of the process are quite involved, however, a section of the process is autocatalyzed by Phosphofructokinase (PFK). This portion of the process is responsible for oscillations in the pathway that leads to the ability of the process to oscillate between an active and an inactive form. In other words, the autocatalytic reactions can turn the process off and on.

Phase Transitions

The initial amounts of reactants determine the distance from chemical equilibrium of the system. The greater the initial concentrations the further the system is from equilibrium. As the initial concentration increases, an abrupt change in order occurs. This abrupt change is known as phase transition. At the phase transition fluctuations in macroscopic quantities, such as chemical concentrations, increase as the system oscillates between the orderly state and the disorderly state. Also, at the phase transition, macroscopic equations, such as the rate equations, fail. Rate equations can be derived from microscopic considerations. The derivations typically rely on a mean field theory approximation to microscopic dynamical equations. Mean field theory breaks down in the presence of large fluctuations. Therefore, since large fluctuations occur in the neighborhood of a phase transition, macroscopic equations, such as rate equations, fail. As the initial concentration increases further, the system settle into an ordered state in which fluctuations are again small.

Autocatalysis

A single chemical reaction is said to have undergone autocatalysis, or be autocatalytic, if the reaction product is itself the catalyst for that reaction.

A *set* of chemical reactions can be said to be "collectively autocatalytic" if a number of those reactions produce, as reaction products, catalysts for enough of the other reactions that the entire set of chemical reactions is self sustaining given an input of energy and food molecules.

Rate law in Autocatalytic Reactions

The rate law for the second order autocatalytic reaction $A + B \rightarrow 2B$ *is* $v = k[A]\ [B]$

The concentrations of A and B vary in time according to:

$$[A] = \frac{[A]_0 + [B]_0}{1 + \dfrac{[B]_0}{[A]_0} e\left([A]_0 + [B]_0\right) kt}$$

$$[B] = \frac{[A]_0 + [B]_0}{1 + \dfrac{[A]_0}{[B]_0} e - \left([A]_0 + [B]_0\right) kt}$$

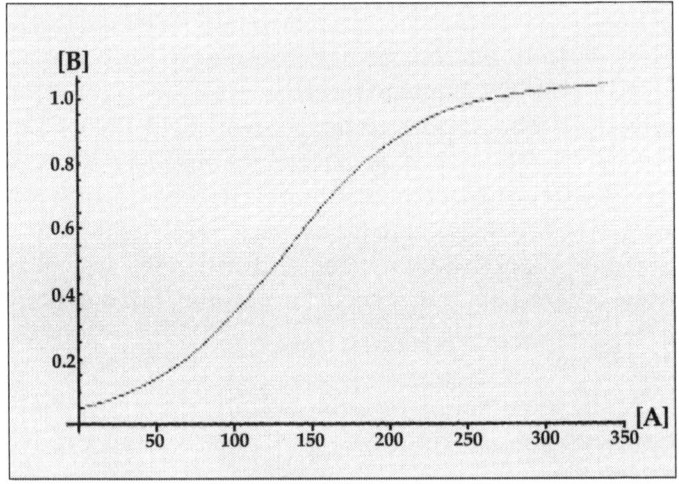

Fig 7.3

Sigmoid variation of product concentration in autocatalytic reactions

The graph for these equations is a sigmoid curve (Fig. 7.3), which is typical for autocatalytic reactions: these chemical reactions proceed slowly at the start because there is little catalyst present, the rate of reaction increases progressively as the reaction proceeds as the amount of catalyst increases and then it again slows down as the reactant concentration decreases. If the concentration of a reactant or product in an experiment follows a sigmoid curve, the reaction is likely to be autocatalytic.

Abiogenesis Hypothesis

British ethologist Richard Dawkins wrote about autocatalysis as a potential explanation for abiogenesis in his 2004 book *The Ancestor's Tale*. He cites experiments performed by Julius Rebek and his colleagues at the Scripps Research Institute in California in which they combined amino adenosine and pentafluorophenyl ester with the autocatalyst amino adenosine triacid ester (AATE). One system from the experiment contained variants of AATE which catalysed the synthesis of themselves. This experiment demonstrated the possibility that autocatalysts could exhibit competition within a population of entities with heredity, which could be interpreted as a rudimentary form of natural selection.

Examples of Autocatalytic Reactions

- Haloform reaction
- Tin pest
- Reaction of Permanganate with Oxalic Acid
- The mechanism of the above reaction
- Vinegar syndrome
- Binding of oxygen by hemoglobin
- The spontaneous degradation of aspirin into salicylic acid and acetic acid, causing very old aspirin in sealed containers to smell mildly of vinegar.
- The α bromination of acetophenone with bromine.

Involvement in Life Processes

Two researchers, Robert Ulanowicz and Stuart Kauffman have suggested that autocatalytic reactions played a central role in the evolution of life, and continue to constitute a basic element in life architecture.

Reaction-diffusion System

Reaction-diffusion systems are mathematical models that describe how the concentration of one or more substances distributed in space changes under the influence of two processes: local chemical reactions in which the substances are converted into each other, and diffusion which causes the substances to spread out in space.

As this description implies, reaction–diffusion systems are naturally applied in chemistry. However, the equation can also describe dynamical processes of non-chemical nature. Examples are found in biology, geology and physics and ecology. Mathematically, reaction–diffusion systems take the form of semi-linear parabolic partial differential equations. They can be represented in the general form

$$\partial tq = \underline{D}\,\nabla^2 q$$

where each component of the vector q(x,*t*) represents the concentration of one substance, $\underline{D}$ is a diagonal matrix of diffusion coefficients, ∇ denotes the Laplace operator and *R* accounts for all local reactions. The solutions of reaction–diffusion equations display a wide range of behaviours, including the formation of travelling waves and wave-like phenomena as well as other self-organized patterns like stripes, hexagons or more intricate structure like dissipative solitons.

One-component Reaction–diffusion Equations

The most simple reaction–diffusion equation concerning the concentration *u* of a single substance in one spatial dimension,

$$\partial_t u = D\partial_x^2 u + R(u)$$

is also referred to as the KPP (Kolmogorov-Petrovsky-Piscounov) equation. If the reaction term vanishes, then the equation represents a pure diffusion process. The corresponding equation is the heat equation. The choice $R(u) = u(1-u)$ yields Fisher's equation that was originally used to describe the spreading of biological populations, the Newell-Whitehead-Segel equation with $R(u) = u(1 - u^2)$ to describe Rayleigh-Benard convection, the more general Zeldovich equation with $R(u) = u(1 - u)(u - a)$ and $0 < a < 1$ that arises in combustion theory, and its particular degenerate case with $R(u) = u^2\ u^3$ that is sometimes referred to as Zeldovich equation as well.

The dynamics of one-component systems is subject to certain restrictions as the evolution equation can also be written in the variational form.M.

$$\partial_t u = -\frac{\partial \Im}{\partial u}$$

and therefore describes a permanent decrease of the "free energy" $\mathfrak{E}$ given by the functional.

$$\mathfrak{E} = \int_{-\infty}^{\infty} \left[\frac{D}{2} \left(\partial_x u \right)^2 + V(u) \right] dx$$

with a potential $V(u)$ such that $R(u)=\mathrm{d}V(u)/\mathrm{d}u$.

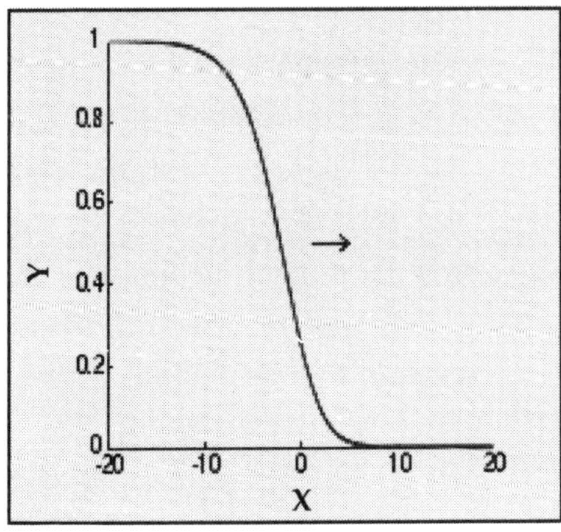

Fig: 7.4

A travelling wave front solution for Fisher's equation

In systems with more than one stationary homogeneous solution, a typical solution is given by travelling fronts connecting the homogeneous states. These solutions move with constant speed without changing their shape and are of the form $u(x, t) = \hat{u}(\xi)$ with $\xi = x - ct$, where c is the speed of the travelling wave. Note that while travelling waves are generically stable structures, all non-monotonous stationary solutions (e.g. localized domains composed of a front-antifront pair) are unstable. For $c = 0$, there is a simple proof for this statement if $u_0(x)$ is a stationary solution and $u=u_0(x) + \ddot{u}(x, t)$ is an infinitesimally perturbed solution, linear stability analysis yields the equation

$$\partial t \tilde{u} = D\partial_x^2 \tilde{u} - U(x)\tilde{u}, \quad U(x) = -R'(u) \mid u=u_0(x)$$

we arrive at the eigenvalue problem

with the ansatz $\ddot{u} = \Psi(x)\exp(-\lambda t)$

$$\hat{H}\Psi = \lambda\Psi, \hat{H} = -D\partial_x^2 + U(x)$$

of Schrödinger type where negative eigenvalues result in the instability of the solution. Due to translational invariance $\Psi = \partial_x u_0(x)$ is a neutral eigenfunction with the eigenvalue $\lambda = 0$, and all other eigenfunctions can be sorted according to an increasing number of knots with the magnitude of the corresponding real eigenvalue increases monotonically with the number of zeros. The eigenfunction $\Psi = \partial_x u_0(x)$ should have at least one zero, and for a non-monotonic stationary solution the corresponding eigenvalue $\lambda = 0$ cannot be the lowest one, thereby implying instability.

To determine the velocity c of a moving front, one may go to a moving coordinate system and look at stationary solutions:

$$D\partial_\xi^2\hat{u}(\xi) + c\partial_\xi\hat{u}(\xi) + R\big(u(\xi)\big) = 0$$

This equation has a nice mechanical analogue as the motion of a mass *D* with position $\hat{u}$ in the course of the "time" ξ under the force *R* with the damping coefficient c which allows for a rather illustrative access to the construction of different types of solutions and the determination of *c*.

When going from one to more space dimensions, a number of statements from one-dimensional systems can still be applied. Planar or curved wave fronts are typical structures, and a new effect arises as the local velocity of a curved front becomes dependent on the local radius of curvature (this can be seen by going to polar coordinates). This phenomenon leads to the so-called curvature-driven instability.

Two-component Reaction-diffusion Equations

Two-component systems allow for a much larger range of possible phenomena than their one-component counterparts. An important idea that was first proposed by Alan Turing is that a state that is stable in the local system should become unstable in the presence of diffusion. This idea seems unintuitive at first glance as diffusion is commonly associated with a stabilizing effect.

A linear stability analysis however shows that when linearizing the general two-component system

$$\begin{pmatrix}\partial_t u \\ \partial_t v\end{pmatrix} = \begin{pmatrix}D_u 0 \\ 0 D_v\end{pmatrix} \begin{pmatrix}\partial_{xx} u \\ \partial_{xx} v\end{pmatrix} + \begin{pmatrix}F(u,v) \\ G(u,v)\end{pmatrix}$$

and perturbing the system against plane waves

$$\tilde{q}_k(x,t) = \begin{pmatrix}u(t) \\ v(t)\end{pmatrix} e^{ik.k}$$

close to a stationary homogeneous solution one finds

$$\begin{pmatrix} \partial_t u_k(t) \\ \partial_t v_k(t) \end{pmatrix} = -k^2 \begin{pmatrix} D_u u_k(t) \\ D_v v_k(t) \end{pmatrix} + R' \begin{pmatrix} u_k(t) \\ v_k(t) \end{pmatrix}.$$

Turing's idea can only be realized in four equivalence classes of systems characterized by the signs of the Jacobian R′ of the reaction function. In particular, if a finite wave vector k is supposed to be the most unstable one, the Jacobian must have the signs

$$\begin{pmatrix} + & - \\ + & - \end{pmatrix}, \begin{pmatrix} + & + \\ - & - \end{pmatrix}, \begin{pmatrix} - & + \\ - & + \end{pmatrix}, \begin{pmatrix} - & - \\ + & + \end{pmatrix}.$$

This class of systems is named *activator-inhibitor system* after its first representative: close to the ground state, one component stimulates the production of both components while the other one inhibits their growth. Its most prominent representative is the FitzHugh–Nagumo equation

$$\partial_t u = \partial_u^2 \Delta u + F(u) - \sigma v$$

$$\tau \partial_t v = \partial_v^2 \Delta v + (u) - v$$

with $f(u) = \lambda_u - u^3 - k$ which describes how an action potential travels through a nerve. Here, d_u, d_v, t, s and λ are positive constants.

When an activator-inhibitor system undergoes a change of parameters, one may pass from conditions under which a homogeneous ground state is stable to conditions under which it is linearly unstable. The corresponding bifurcation may be either a Hopf bifurcation to a globally oscillating homogeneous state with a dominant wave number $k = 0$ or a *Turing bifurcation* to a globally patterned state with a dominant finite wave number. The latter in two spatial dimensions typically leads to stripe or hexagonal patterns.

Three- and more-component reaction–diffusion equations

For a variety of systems, reaction-diffusion equations with more than two components have been proposed, e.g. as models for the Belousov-Zhabotinsky reaction, , for blood clotting or planar gas discharge systems.

While it is known that systems with more components allow for a variety of phenomena not possible in systems with one or two components (e.g. stable running pulses in more than one spatial dimension without global feedback) up to now a systematic overview of the possible phenomena in dependence on the properties of the underlying system is hardly present.

Applications and Universality

In recent times, reaction–diffusion systems have attracted much interest as a prototype model for pattern formation. The above-mentioned patterns (fronts, spirals, targets, hexagons, stripes and dissipative solitons) can be found in various types of reaction-diffusion systems in spite of large discrepancies e.g. in the local reaction terms. It has also been argued that reaction-diffusion processes are an essential basis for processes connected to morphogenesis in biology and may even be related toanimal coats and skin pigmentation.

Another reason for the interest in reaction-diffusion systems is that although they represent nonlinear partial differential equation, there are often possibilities for an analytical treatment.

Experiments

Well-controllable experiments in chemical reaction-diffusion systems have up to now been realized in three ways. First, gel reactors or filled capillary tubes may be used.

Second, temperature pulses on catalytic surfaces have been investigated. Third, the propagation of running nerve pulses is modelled using reaction-diffusion systems.

Aside from these generic examples, it has turned out that under appropriate circumstances electric transport systems like plasmas or semiconductors can be described in a reaction-diffusion approach. For these systems various experiments on pattern formation have been carried out.

Morphogenesis

Morphogenesis (from the Greek *morphê* shape and *genesis* creation, literally, "beginning of the shape"), is the biological process that causes an organism to develop its shape. It is one of three fundamental aspects of developmental biology along with the control of cell growth and cellular differentiation. The process controls the organized spatial distribution of cells during the embryonic development of an organism. Morphogenetic responses may be induced in organisms by hormones, by environmental chemicals ranging from substances produced by other organisms to toxic chemicals or radionuclides released as pollutants, and other plants, or by mechanical stresses induced by spatial patterning of the cells. Morphogenesis can take place in an embyro, a mature organism, in cell culture or inside tumor cell masses.

Morphogenesis also describes the development of unicellular life forms that do not have an embryonic stage in their life cycle, or describes the evolution of a body structure within a taxonomic group.

History

Some of the earliest ideas on how physical and mathematical processes and constraints affect biological growth were written by D'Arcy Wentworth Thompson and Alan Turing. These works postulated the presence of chemical signals and physico-chemical processes such as diffusion, activation, and deactivation in cellular and organismic growth. The fuller understanding of the mechanisms involved in actual organisms required the discovery of DNA and the development of molecular biology and biochemistry.

Molecular Basis

Several types of molecules are particularly important during morphogenesis. Morphogens are soluble molecules that can diffuse and carry signals that control cell differentiation decisions in a concentration-dependent fashion. Morphogens typically act through binding to specific protein receptors. An important class of molecules involved in morphogenesis are transcription factor proteins that determine the fate of cells by interacting with DNA. These can be coded for by master regulatory genes and either activate or deactivate the transcription of other genes; in turn, these secondary gene products can regulate the expression of still other genes in a regulatory cascade. Another class of molecules involved in morphogenesis are molecules that control cell adhesion. For example, during gastrulation, clumps of stem cells switch off their cell-to-cell adhesion, become migratory, and take up new positions within an embryo where they again activate specific cell adhesion proteins and form new tissues and organs. Several examples that illustrate the roles of morphogens, transcription factors and cell adhesion molecules in morphogenesis are discussed below.

Cellular Basis

Morphogenesis arises because of changes in the cellular structure or how cells interact in tissues Certain cell types "sort out". Cell "sorting out" means that when the cells physically interact they move so as to sort into clusters that maximize contact between cells of the same type. The ability of cells to do this comes from differential cell adhesion. Two well-studied types of cells that sort out are epithelial cells and mesenchymal cells. During embryonic development there are some cellular differentiation events during which mesenchymal cells become epithelial cells and at other times epithelial cells differentiate into mesenchymal cells. Following epithelial-mesenchymal transition, cells can migrate away from an epithelium and then associate with other similar cells in a new location.

Adhesion

During embryonic development, cells sort out in different layers due to differential adhesion. Cells that share the same cell-to-cell adhesion molecules separate from cells that have different adhesion molecules. Cells sort based upon differences in adhesion between the cells, so even two populations of cells with different levels of the same adhesion molecule can sort out. In cell culture cells that have the strongest adhesion move to the center of a mixed aggregates of cells.

The molecules responsible for adhesion are called cell adhesion molecules (CAMs). Several types of cell adhesion molecules are known and one major class of these molecules are cadherins. There are dozens of different cadherins that are expressed on different cell types. Cadherins bind to other cadherins in a like-to-like manner: E-cadherin (found on many epithelial cells) binds preferentially to other E-cadherin molecules. Mesenchymal cells usually express other cadherin types such as N-cadherin.

Extracellular Matrix

The extracellular matrix (ECM) is involved with separating tissues, providing structural support or providing a structure for cells to migrate on. Collagen, laminin, and fibronectin are major ECM molecules that are secreted and assembled into sheets, fibers, and gels. Multisubunit transmembrane receptors called integrins are used to bind to the ECM. Integrins bind extracellularly to fibronectin, laminin, or other ECM components, and intracellularly to microfilament-binding proteins a-actinin and talin to link the cytoskeleton with the outside. Integrins also serve as receptors to trigger signal transduction cascades when binding to the ECM. A well-studied example of morphogenesis that involves ECM is mammary gland ductal branching.

8

Catalysis

Catalysis is the process in which the rate of a chemical reaction is either increased or decreased by means of a chemical substance known as a catalyst. Unlike other reagents that participate in the chemical reaction, a catalyst is not consumed by the reaction itself. The catalyst may participate in multiple chemical transformations. Catalysts that speed the reaction are called positive catalysts. Catalysts that slow down the reaction are called negative catalysts or inhibitors. Substances that increase the activity of catalysts are called promoters and substances that deactivate catalysts are called catalytic poisons. For instance, in the reduction of ethyne to ethene, the catalyst is palladium (Pd) partly "poisoned" with lead(II) acetate ($Pb(CH_3COO)_2$). Without the deactivation of the catalyst, the produced ethene will be further reduced to ethane.

The general feature of catalysis is that the catalytic reaction has a lower rate limiting free energy change to the transition state than the corresponding uncatalyzed reaction, resulting in a larger reaction rate at lower temperature. However, the mechanistic origin of catalysis is complex.

Catalysts may affect the reaction environment favorably, e.g. acid catalysts for reactions of carbonyl compounds, form specific intermediates that are not produced naturally, such as osmate esters in osmium tetroxide-catalyzed dihydroxylation of alkenes, or cause lysis of reagents to reactive forms, such as atomic hydrogen in catalytic hydrogenation.

Kinetically, catalytic reactions behave like typical chemical reactions, i.e. the reaction rate depends on the frequency of contact of the reactants in the rate-determining step. Usually, the catalyst participates in this slow step, and rates are limited by amount of catalyst. In heterogeneous catalysis, the diffusion of reagents to the surface and diffusion of products from the surface can be rate determining. Analogous events associated with substrate binding and product dissociation apply to homogeneous catalysts.

Although catalysts are not consumed by the reaction itself, they may be inhibited, deactivated or destroyed by secondary processes. In heterogeneous catalysis, typical secondary processes include coking where the catalyst becomes covered by polymeric side products. Similarly heterogeneous catalysts can dissolve or evaporate into the mobile medium, solution or gas-phase. These side reactions may even include following elementary reaction steps, as in alkali-catalyzed hydrolysis of esters, thus requiring a stoichiometric amount of catalyst.

Background

The production of most industrially important chemicals involves catalysis. Similarly, most biochemically significant processes are catalysed. Research into catalysis is a major field in applied science and involves many areas of chemistry, notably in organometallic chemistry and materials science. Catalysis is relevant to many aspects of environmental science, e.g. the catalytic converter in automobiles and the dynamics of the ozone hole. Catalytic reactions are preferred in environmentally friendly green chemistry due to the reduced amount of waste generated as opposed to stoichiometric reactions in which all reactants are consumed and more side products are formed. The most common catalyst is the proton (H^+). Many transition metals and transition metal complexes are used in catalysis as well. Catalysts called enzymes are important in biology.

A catalyst works by providing an alternative reaction pathway to the reaction product. The rate of the reaction is increased as this alternative route has a lower activation energy than the reaction route not mediated by the catalyst. The disproportionation of hydrogen peroxide to give water and oxygen is a reaction that is strongly affected by catalysts:

$$2H_2O_2 \rightarrow 2\ H_2O + O_2$$

This reaction is favoured in the sense that reaction products are more stable than the starting material, however the uncatalysed reaction is slow. The decomposition of hydrogen peroxide is in fact so slow that hydrogen peroxide solutions are commercially available. Upon the addition of a small amount of manganese dioxide, the hydrogen peroxide rapidly reacts according to the above equation. This effect is readily seen by the effervescence of oxygen. The manganese dioxide may be recovered unchanged, and re-used indefinitely, and thus is not consumed in the reaction. Accordingly, manganese dioxide *catalyses* this reaction.

General Principles of Catalysis

Typical Mechanism

Catalysts generally react with one or more reactants to form intermediates that subsequently give the final reaction product, in the process regenerating the catalyst. The following is a typical reaction scheme, where *C* represents the catalyst, X and Y are reactants, and Z is the product of the reaction of X and Y:

$X + C \rightarrow XC$

$Y + XC \rightarrow XYC$

$XYC \rightarrow CZ$

$CZ \rightarrow C + Z$

Although the catalyst is consumed by reaction 1, it is subsequently produced by reaction 4, so for the overall reaction:

$X + Y \rightarrow Z$

As a catalyst is regenerated in a reaction, often only small amounts are needed to increase the rate of the reaction. In practice, however, catalysts are sometimes consumed in secondary processes.

As an example of this process, in 2008 Danish researchers first revealed the sequence of events when oxygen and hydrogen combine on the surface of titanium dioxide (TiO_2, or *titania*) to produce water. With a time-lapse series of scanning tunneling microscopy images, they determined the molecules undergo adsorption, dissociation and diffusion before reacting. The intermediate reaction states were: HO_2, H_2O_2, then H_3O_2 and the final reaction product (water molecule dimers), after which the water molecule desorbs from the catalyst surface.

Catalysts work by providing an (alternative) mechanism involving a different transition state and lower activation energy. Consequently, more molecular collisions have the energy needed to reach the transition state. Hence, catalysts can enable reactions that would otherwise be blocked or slowed by a kinetic barrier. The catalyst may increase reaction rate or selectivity, or enable the reaction at lower temperatures. This effect can be illustrated with a Boltzmann distribution and energy profile diagram.

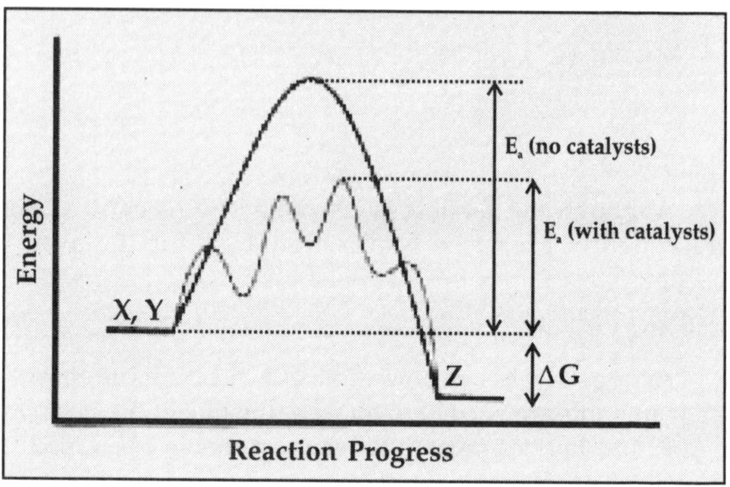

Fig. 8.1

Genetic potential energy diagram showing the effect of a catalyst in a hypothetical exothermic chemical reaction X + Y to give Z. The presence of the catalyst opens a different reaction pathway with a lower activation energy. The final result and the overall thermodynamics are the same.

Catalysts do not change the extent of a reaction: they have no effect on the chemical equilibrium of a reaction because the rate of both the forward and the reverse reaction are both affected (see also thermodynamics). The fact that a catalyst does not change the equilibrium is a consequence of the second law of thermodynamics. Suppose there was such a catalyst that shifted an equilibrium. Introducing the catalyst to the system would result in reaction to move to the new equilibrium, producing energy. Production of energy is a necessary result since reactions are spontaneous if and only if Gibbs free energy is produced, and if there is no energy barrier, there is no need for catalyst. Then, removing the catalyst would also result in reaction, producing energy; i.e. the addition and its reverse process, removal, would both produce energy. Thus, a catalyst that could change the equilibrium would be a perpetual motion machine, a contradiction to the laws of thermodynamics.

If a catalyst does change the equilibrium, then it must be consumed as the reaction proceeds, and thus it is also a reactant. Illustrative is the base-catalysed hydrolysis of esters.

The SI derived unit for measuring the catalytic activity of a catalyst is the katal, which is moles per second. The activity of a catalyst can also be described by the turn over number (or TON) and the catalytic efficiency by the *turn over frequency* (TOF). The biochemical equivalent is the enzyme unit. For more information on the efficiency of enzymatic catalysis, see the article on Enzymes.

The catalyst stabilizes the transition state more than it stabilizes the starting material. It decreases the kinetic barrier by decreasing the *difference* in energy between starting material and transition state.

Typical Catalytic Materials

The chemical nature of catalysts is as diverse as catalysis itself, although some generalizations can be made. Proton acids are probably the most widely used catalysts, especially for the many reactions involving water, including hydrolysis and its reverse. Multifunctional solids often are catalytically active, e.g. zeolites, alumina and certain forms of graphitic carbon. Transition metals are often used to catalyse redox reactions (oxidation, hydrogenation). Many catalytic processes, especially those involving hydrogen, require platinum metals.

Some so-called catalysts are really "precatalysts." Precatalysts convert to catalysts in the reaction. For example, Wilkinson's catalyst $RhCl(PPh_3)_3$ loses one triphenylphosphine ligand before entering the true catalytic cycle. Precatalysts are easier to store but are easily activated in situ. Because of this preactivation step, many catalytic reactions involve an induction period.

Types of Catalysis

Catalysts can be either heterogeneous or homogeneous, depending on whether a catalyst exists in the same phase as the substrate. Biocatalysts are often seen as a separate group.

Heterogeneous Catalysts

Heterogeneous catalysts are those which act in a different phases than the reactants. Most heterogeneous catalysts are solids that act on substrates in a liquid or gaseous reaction mixture. Diverse mechanisms for reactions on surfaces are known, depending on how the adsorption takes place (Langmuir-Hinshelwood and Eley-Rideal).

For example, in the Haber process, finely divided iron serves as a catalyst for the synthesis of ammonia from nitrogen and hydrogen. The reacting gases adsorb onto "active sites" on the iron particles. Once adsorbed, the bonds within the reacting molecules are weakened, and new bonds between the resulting fragments form in part due to their close proximity. In this way the particularly strong triple bond in nitrogen is weakened and the hydrogen and nitrogen atoms combine faster than would be the case in the gas phase, so the rate of reaction increases.

Heterogeneous catalysts are typically "supported," which means that the catalyst is dispersed on a second material that enhances the effectiveness or minimizes their cost. Sometimes the support is merely a surface upon which the catalyst is spread to increase the surface area. More often, the support and the catalyst interact, affecting the catalytic reaction.

Homogeneous Catalysts

Homogeneous catalysts function in the same phase as the reactants, but the mechanistic principles invoked in heterogeneous catalysis are generally applicable. Typically homogeneous catalysts are dissolved in a solvent with the substrates. One example of homogeneous catalysis involves the influence of H+ on the esterification of esters, e.g. methyl acetate from acetic acid and methanol For inorganic chemists, homogeneous catalysis is often synonymous with organometallic catalysts.

Electrocatalysts

In the context of electrochemistry, specifically in fuel cell engineering, various metal-containing catalysts are used to enhance the rates of the half reactions that comprise the fuel cell. One common type of fuel cell electrocatalyst is based upon nanoparticles of platinum that are supported on slightly larger carbon particles. When this platinum electrocatalyst is in contact with one of the electrodes in a fuel cell, it increases the rate of oxygen reduction to water (or hydroxide or hydrogen peroxide).

Organocatalysis

Whereas transition metals sometimes attract most of the attention in the study of catalysis, organic molecules without metals can also possess catalytic properties. Typically, organic catalysts require a higher loading (or amount of catalyst per unit amount of reactant) than transition metal-based catalysts, but these catalysts are usually commercially available in bulk, helping to reduce costs. In the early 2000s, organocatalysts were considered "new generation" and are competitive to traditional metal-containing catalysts. Enzymatic reactions operate via the principles of organic catalysis.

Significance of Catalysis

Estimates are that 90% of all commercially produced chemical products involve catalysts at some stage in the process of their manufacture In 2005, catalytic processes generated about $900 billion in products worldwide. Catalysis is so pervasive that subareas are not readily classified. Some areas of particular concentration are surveyed below.

Energy Processing

Petroleum refining makes intensive use of catalysis for alkylation, catalytic cracking (breaking long-chain hydrocarbons into smaller pieces), naphtha reforming and steam reforming (conversion of hydrocarbons into synthesis gas). Even the exhaust from the burning of fossil fuels is treated via catalysis: Catalytic converters, typically composed of platinum and rhodium, break down some of the more harmful byproducts of automobile exhaust.

$$2\,CO + 2\,NO \rightarrow 2\,CO_2 + N_2$$

With regards to synthetic fuels, an old but still important process is the Fischer-Tropsch synthesis of hydrocarbons from synthesis gas, which itself is processed via water-gas shift reactions, catalysed by iron. Biodiesel and related biofuels require processing via both inorganic and biocatalysts.

Fuel cells rely on catalysts for both the anodic and cathodic reactions.

Bulk Chemicals

Some of the largest-scale chemicals are produced via catalytic oxidation, often using oxygen. Examples include nitric acid (from ammonia), sulfuric acid (from sulfur dioxide to sulfur trioxide by the chamber process), terephthalic acid from p-xylene, and acrylonitrile from propane and ammonia. Many other chemical products are generated by large-scale reduction, often via hydrogenation. The largest-scale example is ammonia, which is prepared via the Haber process from nitrogen. Methanol is prepared from carbon monoxide. Bulk polymers derived from ethylene and propylene are often prepared via Ziegler-Natta catalysis. Polyesters, polyamides, and isocyanates are derived via acid-base catalysis. Most carbonylation processes require metal catalysts, examples include the Monsanto acetic acid process and hydroformylation.

Fine Chemicals

Many fine chemicals are prepared via catalysis; methods include those of heavy industry as well as more specialized processes that would be prohibitively expensive on a large scale. Examples include olefin metathesis using Grubbs' catalyst, the Heck reaction, and Friedel-Crafts reactions. Because most bioactive compounds are chiral, many pharmaceuticals are produced by enantioselective catalysis.

Food Processing

One of the most obvious applications of catalysis is the hydrogenation (reaction with hydrogen gas) of fats using nickel catalyst to produce margarine Many other foodstuffs are prepared via biocatalysis.

Biology

In nature, enzymes are catalysts in metabolism and catabolism. Most biocatalysts are protein-based, i.e. enzymes, but other classes of biomolecules also exhibit catalytic properties including ribozymes, and synthetic deoxyribozymes. Biocatalysts can be thought of as intermediate between homogenous and heterogeneous catalysts, although strictly speaking soluble enzymes are homogeneous catalysts and membrane-bound enzymes are heterogeneous. Several factors affect the activity of enzymes (and other catalysts) including temperature, pH, concentration of enzyme, substrate, and products. A particularly important reagent in enzymatic reactions is water, which is the product of many bond-forming reactions and a reactant in many bond-breaking processes. Enzymes are employed to prepare many commodity chemicals including high-fructose corn syrup and acrylamide.

In the Environment

Catalysis impacts the environment by increasing the efficiency of industrial processes, but catalysis also plays a direct role in the environment. A notable example is the catalytic role of Chlorine free radicals in the breakdown of ozone. These radicals are formed by the activation of ultraviolet radiation on chlorofluorocarbons (CFCs).

$$Cl + O3 \quad ClO^{\cdot} + O_2$$

$$ClO + O \quad Cl^{\cdot} + O_2$$

History

In a general sense, anything that increases the rate of a process is a "catalyst", a term derived from Greek , meaning "to annul," or "to untie," or "to pick up." The phrase *catalysed processes* was coined by Jöns Jakob Berzelius in 1836 to describe reactions that are accelerated by substances that remain unchanged after the reaction. Other early chemists involved in catalysis were Alexander Mitscherlich who referred to *contact processes* and Johann Wolfgang Döbereiner who spoke of *contact action* and whose lighter based on hydrogen and a platinum sponge became a huge commercial success in the 1820s. Humphry Davy discovered the use of platinum in catalysis. In the 1880s, Wilhelm Ostwald at Leipzig University started a systematic investigation into reactions that were catalyzed by the presence of acids and bases, and found that chemical reactions occur at finite rates and that these rates can be used to determine the strengths of acids and bases. For this work, Ostwald was awarded the 1909 Nobel Prize in Chemistry.

9

Rate Equation

The rate law or rate equation for a chemical reaction is an equation which links the reaction rate with concentrations or pressures of reactants and constant parameters (normally rate coefficients and partial reaction orders). To determine the rate equation for a particular system one combines the reaction rate with a mass balance for the system. For a generic reaction A + B   C the *simple rate equation* (as opposed to the much more common *complicated rate equations*) is of the form:

$$r = k[A]^m[B]^n$$

where [A] and [B] express the concentration of the species A and B, respectively (usually in moles per liter (molarity), m and n are not the respective stoichiometric coefficients of the balanced equation; they must be determined experimentally. k is the *rate coefficient* or *rate constant* of the reaction. The value of this coefficient k depends on conditions such as temperature, ionic strength, surface area of the adsorbent or light irradiation. For elementary reactions, the rate equation can be derived from first principles using collision theory. Again, m and n are NOT always derived from the balanced equation.

The rate equation of a reaction with a multi-step mechanism cannot, in general, be deduced from the stoichiometric coefficients of the overall reaction; it must be determined experimentally. The equation may involved fractional exponential coefficients or it may depend on the concentration of an intermediate species.

The rate equation is a differential equation, and it can be integrated in order to obtain an integrated rate equation that links concentrations of reactants or products with time.

If the concentration of one of the reactants remains constant (because it is a catalyst or it is in great excess with respect to the other reactants) its concentration can be included in the rate constant, obtaining a pseudo constant: if B is the reactant whose concentration is constant then $r = k[A][B] = k'[A]$. The second order rate equation has been reduced to a pseudo first order rate equation. This makes the treatment to obtain an integrated rate equation much easier.

Zero-order Reactions

A Zero-order reaction has a rate which is independent of the concentration of the reactant(s). Increasing the concentration of the reacting species will not speed up the rate of the reaction. Zero-order reactions are typically found when a material that is required for the reaction to proceed, such as a surface or a catalyst, is saturated by the reactants. The rate law for a zero-order reaction is

$$r = k$$

where r is the reaction rate, and k is the reaction rate coefficient with units of concentration/time. If, and only if, this zero-order reaction:

1. occurs in a closed system;
2. there is no net build-up of intermediates; and
3. there are no other reactions occurring, it can be shown by solving a Mass balance for the system that:

$$r = \frac{d[A]}{dt} = k$$

If this differential equation is integrated it gives an equation which is often called the *integrated zero-order rate law*

$$[A]_t = -kt + [A]_0$$

where $[A]_t$ represents the concentration of the chemical of interest at a particular time, and $[A]_0$ represents the initial concentration.

A reaction is zero order if concentration data are plotted versus time and the result is a straight line. The slope of this resulting line is the negative of the zero order rate constant k.

The half-life of a reaction describes the time needed for half of the reactant to be depleted (same as the half-life involved in nuclear decay, which is a first-order reaction). For a zero-order reaction the half-life is given by

$$t_{\frac{1}{2}} = \frac{[A]_0}{2k}$$

Example of a Zero-order reaction

Reversed harber process : $2\ NH_3\ (g) \rightarrow 3\ H_2(g) + N_2\ (g)$

It should be noted that the order of a reaction cannot be deduced from the chemical equation of the reaction.

First-order Reactions

A first-order reaction depends on the concentration of only one reactant (*a unimolecular reaction*). Other reactants can be present, but each will be zero-order. The rate law for an elementary reaction that is first order with respect to a reactant A is:

$$r = \frac{d[A]}{dt} = k[A]$$

k is the first order rate constant, which has units of 1/time.

The integrated first-order rate law is

$\ln [A] = -kt + \ln [A]_0$

A plot of $\ln[A]$ vs. time t gives a straight line with a slope of - k.

The half life of a first-order reaction is independent of the starting concentration and is given by .

$$t_{\frac{1}{2}} = \frac{\ln(2)}{k}$$

Examples of reactions that are first-order with respect to the reactant:

$$H_2O_2(l) \rightarrow H_2O(l) + \frac{1}{2}O_2(g)$$

$$SO_2Cl_2(l) \rightarrow S_2O(g) + Cl_2(g)$$

$$2N_2O_5(g) \rightarrow 4NO_2(g) + O_2(g)$$

Second-order Reactions

A second-order reaction depends on the concentrations of one second-order reactant, or two first-order reactants.

For a second order reaction, its reaction rate is given by:

$r = k[A]^2$ or $r = k\ [A]\ [B]$ or $r = k\ [B]^2$

The *integrated second-order* rate laws are respectively

$$\frac{1}{[A]} = kt + \frac{1}{[A]_0} \text{ or}$$

$$\frac{[A]}{[B]} = \frac{[A]_0}{[B]_0} e^{([A]0-[B]0)kt}$$

$[A]_0$ and $[B]_0$ must be different, in order to obtain that integrated equation.

The half-life equation for a second-order reaction dependent on one second-order reactant is

$$t_{\frac{1}{2}} = \frac{1}{k[A]_0}$$

For a second-order reaction half-lives progressively double.

Another way to present the above rate laws is to take the log of both sides: $\ln r - \ln k + 2\ln[A]$

Examples of a Second-order reaction

$2NO_2(g) \rightarrow 2NO(g) + O_2(g)$

Pseudo First Order

Measuring a second order reaction rate can be problematic: the concentrations of the two reactants must be followed simultaneously, which is more difficult; or measure one of them and calculate the other as a difference, which is less precise. A common solution for that problem is the *pseudo first order approximation.*

If *either* [A] or [B] remain constant as the reaction proceeds, then the reaction can be considered pseudo first order because in fact it only depends on the concentration of one reactant. If for example [B] remains constant then:

$$r = k\,[A]\,[B] = k'\,[A]$$

where $k' = k[B]_0$ (k' or k_{obs} with units s^{-1}) and we have an expression identical to the first order expression above.

One way to obtain a pseudo first order reaction is to use a large excess of one of the reactants ([B]>>[A] would work for the previous example) so that, as the reaction progresses only a small amount of the reactant is consumed and its concentration can be considered to stay constant. By collecting k' for many reactions with different (but excess) concentrations of [B]; a plot of k' versus [B] gives k (the regular second order rate constant) as the slope.

Equilibrium Reactions or Opposed Reactions

A pair of forward and reverse reactions may define an equilibrium process. For example A and B react into X and Y and vice versa (s, t, u and v are the stoichiometric coefficients):

$$sA + tB \rightleftharpoons uX + vY$$

The reaction rate expression for the above reactions (assuming they each are elementary) can be expressed as:

$$r = k_1[A]^S[B]^t - k_2[X]^u[Y]^v$$

where: k_1 is the rate coefficient for the reaction which consumes A and B; k_2 is the rate coefficient for the backwards reaction, which consumes X and Y and produces A and B.

The constants k_1 and k_2 are related to the equilibrium coefficient for the reaction (K) by the following relationship (set r=0 in balance):

$$k_1[A]^s[B]^t = k_2[X]^u[Y]^v$$

$$K = \frac{[X]^u[Y]^v}{[A]^s[B]^t} = \frac{k_1}{k_2}$$

In a simple equilibrium between two species:

$A \rightleftharpoons B$

the constant K at equilibrium is expressed as:

$$K \overset{\text{def}}{=} \frac{k_f}{k_b} = \frac{[B]_e}{[A]_e}$$

When the concentration of A at equilibrium is that of the concentration at time 0 minus the conversion in moles

$$[A]_e = [A]_0 - x$$

with x equal to the concentration of B at equilibrium

$$[B]_e = x$$

then it follows that

$$[B]_e = x = \frac{k_f}{k_f + k_b}[A]_0$$

$$[A]_e = [A]_0 - x = \frac{k_b}{k_f + k_b}[A]_0$$

The reaction rate becomes:

$$\frac{dx}{dt} = \frac{kf[A]_0}{x_e}(x_e - x)$$

which results in

$$ln\left(\frac{[A]_0 - [A]_e}{[A_t] - [A]_e}\right) = \left(k_f + k_b\right)t$$

A plot of the negative natural logarithm of the concentration of A in time minus the concentration at equilibrium versus time t gives a straight line with slope $k_f + k_b$. By measurement of A_e and B_e the values of K and the two reaction rate constants will be known.

When the equilibrium constant is close to unity and the reaction rates very fast for instance in conformational analysis of molecules, other methods are required for the determination of rate constants for instance by complete lineshape analysis in NMR spectroscopy.

Consecutive Reactions

If the rate constants for the following reaction are k_1 and k_2; $A \rightarrow B \rightarrow C$ then the rate equation is:

For reactant A: $\frac{d[A]}{dt} = -k_1[A]$

For reactant B: $\frac{d[B]}{dt} = k_1[A] - k_2[B]$

For product C: $\frac{d[C]}{dt} = k_2[B]$

These differential equations can be solved analytically and the integrated rate equations (supposing that initial concentrations of every substance except A are zero) are

$$[A] = [A]_0 e^{-k_1 t}$$

$$[B] = [A]_0 \frac{k1}{k2-k1}\left(e^{-k_1 t} - e^{-k_2 t}\right)$$

$$[C] = \frac{[A]_0}{k_2 - k_1}\left[k_2\left(1-e^{-k_1 t}\right)-k_1\left(1-e^{-k_2 t}\right)\right]$$

$$= [A]_0\left(1 + \frac{k_1 e^{-k_2 t} - k_2 e^{-k1t}}{k_2 - k_1}\right)$$

The steady state approximation leads to very similar results in an easier way.

Parallel or Competitive Reactions

When a substance reacts simultaneously to give two different products, a parallel or competitive reaction is said to take place.

- Two first order reactions:

A $\rightarrow$ B and A $\rightarrow$ C, with constants k_1 and k_2 and rate equations $\frac{d[A]}{dt} = (k_1 + k_2)[A] \frac{d[B]}{dt} = k_1[A]$, and $\frac{d[C]}{dt} = k_2[A]$

The integrated rate equations are then

$$[A] = [A]_0 e^{-(k_1+k_2)t}$$

$$[B] = \frac{k_1}{k_1+k_2}[A]_0\left(1-e^{-(k_1+k_2)^t}\right); \text{ and}$$

$$[C] = \frac{k_1}{k_1+k_2}[A]_0\left(1-e^{-(k_1+k_2)^t}\right)$$

One important relationship in this case is $\frac{[B]}{[C]} = \frac{k_1}{k_2}$

One first order and one second order reaction: This can be the case when studying a bimolecular reaction and a simultaneous hydrolysis (which can be treated as pseudo order one) takes place: the hydrolysis complicates the study of the reaction kinetics, because some reactant is being "spent" in a parallel reaction. For example A reacts with R to give our product C, but meanwhile the hydrolysis reaction takes away an amount of A to give B, a byproduct: A + H_2O → B and A + R → C. The rate equations are and

$$\frac{d[B]}{dt} = k1[A][H_2O] = k'_1[A].$$

Where k_1 is the pseudo first order constant.

The integrated rate equation for the main product [C] is $[C] = [R]_0\left[1 - e^{-\frac{k_2}{k'_1}[A]_0\left(1-e^{-k'_1 t}\right)}\right]$

which is equivalent to $ln\frac{[R]0}{[R]0-[C]} = \frac{k2[A]0}{k'1}\left(1-e^{-k'_1 t}\right)$.

Concentration of B is related to that of C through

$$[B] = -\frac{k'_1}{k_2} in\left(1 - \frac{[C]}{[R]_0}\right)$$

The integrated equations were analytically obtained but during the process it was assumed that $[A]_0 - [C] \approx [A]_0$ therefore, previous equation for [C] can only be used for low concentrations of [C] compared to $[A]_0$.

Reaction Rate

The reaction rate or rate of reaction for a reactant or product in a particular reaction is intuitively defined as how fast a reaction takes place. For example, the oxidation of iron under the atmosphere is a slow reaction which can take many years, but the combustion of butanc in a fire is a reaction that takes place in fractions of a second.

Chemical kinetics is the part of physical chemistry that studies reaction rates. The concepts of chemical kinetics are applied in many disciplines, such as chemical engineering, enzymology and environmental engineering.

Formal Definition of Reaction Rate

Consider a typical chemical reaction:

$$aA + bB \rightarrow pP + qQ$$

The lowercase letters (a, b, p, and q) represent stoichiometric coefficients, while the capital letters represent the reactants (A and B) and the products (P and Q).

According to Jerrica IUPAC's Gold Book definition the reaction rate *v* (also *r* or *R*) for a chemical reaction occurring in a closed system under constant-volume conditions, without a build-up of reaction intermediates, is defined as:

$$v = -\frac{1}{a}\frac{d[A]}{dt} = \frac{1}{b}\frac{d[B]}{dt} = \frac{1}{p}\frac{d[P]}{dt} = \frac{1}{q}\frac{d[Q]}{dt}$$

The IUPAC recommends that the unit of time should always be the second. In such a case the rate of reaction differs from the rate of increase of concentration of a product P by a constant factor (the reciprocal of its stoichiometric number) and for a reactant A by minus the reciprocal of the stoichiometric number. Reaction rate usually has the units of mol dm^{-3} s^{-1}. It is important to bear in mind that the previous definition is only valid for a *single reaction*, in a *closed system* of *constant volume*. This most usually implicit assumption must be stated explicitly, otherwise the definition is incorrect: If water is added to a pot containing salty water, the concentration of salt decreases, although there is no chemical reaction.

For any system in general the full mass balance must be taken into account: In - Out + Generation = Accumulation

$$F_{A0} - FA = \int_0^v r dV = \frac{dN_A}{dt}$$

When applied to the simple case stated previously this equation reduces to:

$$v = \frac{dA}{dt}$$

For a single reaction in a closed system of varying volume the so called *rate of conversion* can be is used, in order to avoid handling concentrations. It is defined as the derivative of the extent of reaction with respect to time.

$$\xi = \frac{d\xi}{dt} = \frac{1}{v_i}\frac{dn_i}{d_t} = \frac{1}{v_i}\left(v\frac{dCi}{dt} + C_i\frac{dV}{dt}\right)$$

is the stoichiometric coefficient for substance *i* , *v* is the volume of reaction and C is the concentration of substance *i*.

When side products reaction intermediates are formed, the IUPAC recommends the use of the terms *rate of appearance* and *rate of disappearance* for products and reactants, respectively.

Reaction rates may also be defined on a basis that is not the volume of the reactor. When a catalyst is used the reaction rate may be stated on a catalyst weight (mol g^{-1} s^{-1}) or surface area (mol m-2 s-1) basis. If the basis is a specific catalyst site that may be rigorously counted by a specified method, the rate is given in units of s-1 and is called a turnover frequency.

Factors influencing Rate of Reaction

- *Concentration*: Reaction rate increases with concentration, as described by the rate law and explained by collision theory. As reactant concentration increases, the frequency of collision increases.

- *The nature of the reaction*: Some reactions are naturally faster than others. The number of reacting species, their physical state (the particles that form solids move much more slowly than those of gases or those in solution), the complexity of the reaction and other factors can influence greatly the rate of a reaction.
- *Temperature*: Usually conducting a reaction at a higher temperature delivers more energy into the system and increases the reaction rate by causing more collisions between particles, as explained by collision theory. However, the main reason why it increases the rate of reaction is that more of the colliding particles will have the necessary activation energy resulting in more successful collisions (when bonds are formed between reactants). The influence of temperature is described by the Arrhenius equation. As a rule of thumb, reaction rates for many reactions double for every 10 degrees Celsius increase in temperature, though the effect of temperature may be very much larger or smaller than this (to the extent that reaction rates can be independent of temperature or decrease with increasing temperature.)

For example, coal burns in a fireplace in the presence of oxygen but it doesn't when it is stored at room temperature. The reaction is spontaneous at low and high temperatures but at room temperature its rate is so slow that it is negligible. The increase in temperature, as created by a match, allows the reaction to start and then it heats itself, because it is exothermic. That is valid for many other fuels, such as methane, butane, hydrogen.

- *Solvent*: Many reactions take place in solution and the properties of the solvent affect the reaction rate. The ionic strength also has an effect on reaction rate.

- *Pressure*: The rate of gaseous reactions increases with pressure, which is, in fact, equivalent to an increase in concentration of the gas. For condensed-phase reactions, the pressure dependence is weak.

- *Electromagnetic Radiation*: Electromagnetic radiation is a form of energy. As such, it may speed up the rate or even make a reaction spontaneous as it provides the particles of the reactants with more energy. This energy is in one way or another stored in the reacting particles (it may break bonds, promote molecules to electronically or vibrationally excited states...) creating intermediate species that react easily.

For example when methane reacts with chlorine in the dark, the reaction rate is very slow. It can be sped up when the mixture is put under diffused light. In bright sunlight, the reaction is explosive.

- *A catalyst*: The presence of a catalyst increases the reaction rate e(in both the forward and reverse reactions) by providing an alternative pathway with a lower activation energy.

For example, platinum catalyzes the combustion of hydrogen with oxygen at room temperature.

- *Isotopes*: The kinetic isotope effect consists in a different reaction rate for the same molecule if it has different isotopes, usually hydrogen isotopes, because of the mass difference between hydrogen and deuterium.

- *Surface Area*: In reactions on surfaces, which take place for example during heterogeneous catalysis, the rate of reaction increases as the surface area does. That is due to the fact that more particles of the solid are exposed and can be hit by reactant molecules.

- *Order*: The order of the reaction controls how the reactant concentration affects reaction rate.

- *Stirring*: Stirring can have a strong effect on the rate of reaction for heterogeneous reactions.
- *Intensity of light*: The reactants involved in a photochemical reaction absorb energy from light and other EM radiation. As the intensity of light increases, the particles absorb more energy. Thus their kinetic energy increases, and there are more productive collisions. Hence the rate of reaction increases. The converse is also true as light intensity decreases.

All the factors that affect a reaction rate are taken into account in the rate equation of the reaction.

Rate Equation

For a chemical reaction n A + m B = C + D, the rate equation or rate law is a mathematical expression used in chemical kinetics to link the rate of a reaction to the concentration of each reactant. It is of the kind:

$$r = k(T)[A]^{n'} [B]^{m'}$$

In this equation k(T) is the *reaction rate coefficient* or *rate constant*, although it is not really a constant, because it includes all the parameters that affect reaction rate, except for concentration, which is explicitly taken into account. Of all the parameters described before, temperature is normally the most important one.

The exponents n' and m' are called reaction orders and depend on the reaction mechanism.

Stoichiometry, molecularity (the actual number of molecules colliding) and reaction order only coincide necessarily in elementary reactions, that is, those reactions that take place in just one step. The reaction equation for elementary reactions coincides with the process taking place at the atomic level, i.e. n molecules of type A are colliding with m molecules of type B (n plus m is the molecularity).

For gases the rate law can also be expressed in pressure units using e.g. the ideal gas law.

By combining the rate law with a mass balance for the system in which the reaction occurs, an expression for the rate of change in concentration can be derived. For a closed system with constant volume such an expression can look like

$$\frac{d[C]}{dt} = k(T)[A]^{n'}[B]^{m'}$$

Temperature Dependence

Each reaction rate coefficient k has a temperature dependency, which is usually given by the Arrhenius equation:

$$k = Ae - \frac{E_q}{RT}$$

E_a is is the activation energy and R is the gas constant. Since at temperature T the molecules have energies given by a Boltzmann distribution, one can expect the number of collisions with energy greater than E_a to be proportional to

$$e\frac{-E_q}{RT}$$

A is the pre-exponential factor or frequency factor.

The values for A and E_a are dependent on the reaction. There are also more complex equations possible, which describe temperature dependence of other rate constants which do not follow this pattern.

Pressure Dependence

The pressure dependence of the rate constant for condensed-phase reactions (i.e., when reactants and products are solids or liquid) is usually sufficiently weak in the range of pressures normally encountered in industry that it is neglected in practice.

The pressure dependence of the rate constant is associated with the activation volume. For the reaction proceeding through an activation-state complex:

$$A + B \rightleftharpoons [A...B]^t \rightarrow P$$

the activation volume, $\Delta V^{\neq}$ is

$$\Delta V^t = \overline{V}_{\neq} - \overline{V}_A - \overline{V}_B$$

where $\overline{V}$ denote the partial molar volumes of the reactants and products and $\neq$ indicates the activation-state complex.

For the above reaction, one can expect the change of the reaction rate constant (based either on mole-fraction or molal-concentration) with pressure at constant temperature to be:

$$-RT\left(\frac{\partial \ln k_x}{\partial P}\right)_T = \Delta V^{\neq}$$

In practice, the matter can be complicated because the partial molar volumes and the activation volume can themselves be a function of pressure.

Reactions can increase or decrease their rates with pressure, depeding on the value of $\Delta V^{\neq}$. As an example of the possible magnitude of the pressure effect, some organic reactions were shown to double the reaction rate when the pressure was increased from atomospheric (0.1 MPa) to 50 MPa (which gives $\Delta V^{\neq}$ = -0.025 L/mol).

Examples

For the reaction

$$2H_2(g) + 2\ NO\ (g) \rightarrow N_2\ (g) + 2H_2O\ (g)$$

The rate equation is

$$r = k[H_2]^1\ [NO]^2$$

The rate equation does not simply reflect the reactants stoichiometric coefficients in the overall reaction: it is first order in H_2, although the stoichiometric coefficient is 2 and it is second order in NO.

In chemical kinetics the overall reaction is usually proposed to occur through a number of elementary steps. Not all of these steps affect the rate of reaction; normally it is only the slowest elementary step that affect the reation rate. For example, in:

$$2NO \rightleftharpoons N_2O_2 \text{ (fast equilibrium)}$$

$$N_2O_2 + H_2 \rightarrow N_2O + H_2) \text{ (Slow)}$$

$$N_2O + H_2 \rightarrow N_2 + H_2) \text{ (fast)}$$

Reactions 1 and 3 are very rapid compared to the second, so it is the slowest reaction that is reflected in the rate equation. The slow step is considered the rate determining step. The orders of the rate equation are those from the rate determining step.

10

Collision Theory

The Collision theory, proposed by Max Trautz and William Lewis in 1916 and 1918, qualitatively explains how chemical reactions occur and why reaction rates differ for different reactions. This theory is based on the idea that reactant particles must collide for a reaction to occur, but only a certain fraction of the total collisions have the energy to connect effectively and cause the reactants to transform into products. This is because only a portion of the molecules have enough energy and the right orientation (or "angle") at the moment of impact to break any existing bonds and form new ones. The minimal amount of energy needed for this to occur is known as activation energy. Particles from different elements react with each other by releasing activation energy as they hit each other. If the elements react with each other, the collision is called successful, but if the concentration of at least one of the elements is too low, there will be fewer particles for the other elements to react with and the reaction will happen much more slowly.

Collision theory is closely related to chemical kinetics.

Rate Constant

The rate constant for a bimolecular gas phase reaction, as predicted by collision theory is:

$$k(T) = Z\ \rho \exp\ \rho\left(\frac{-E_\alpha}{RT}\right)$$

where

Z is the collision frequency.

P is the steric factor.

Ea is the activation energy of the reaction.

T is the temperature.

R is gas constant.

And the collision frequency is:

$$Z = N_A^2 \sigma AB \sqrt{\frac{8k_B T}{\pi\mu_{AB}}}$$

where

N_A is Avogadro's number

s_{AB} is the reaction cross section

k_B is Boltzmann's constant

μ_{AB} is the reduced mass of the reactants

Quantitative Insights

Derivation

Collision theory can only be applied quantitatively to bimolecular reactions, of the kind

A + B →

In collision theory it is considered that two particles A and B will collide if their nuclei get closer than a certain distance. The area around a molecule A in which it can collide with an approaching B molecule is called the cross section (σ_{AB}) of the reaction and is, in principle, the area corresponding to a circle whose radius (r_{AB}) is the sum of the radii of both reacting molecules, which are supposed to be spherical. A moving molecule will therefore sweep a volume $\pi r^2{}_{AB} C_A$ per second as it moves, where is the average velocity of the particle.

From kinetic theory it is known that a molecule of A has an average velocity (different from root mean square velocity) of $C_A = \sqrt{\frac{8k_B T}{\pi m_A}}$,

where is Boltzmann constant and is the mass of the molecule.

The solution of the two body problem states that two different moving bodies can be treated as one body which has the reduced mass of both and moves with the velocity of the center of mass, so, in this system μ_{AB} must be used instead of m_A.

Therefore, the total collision frequency of all A molecules, with all B molecules, is:

$$N_A^2 \sigma_{AB} \sqrt{\frac{8k_B T}{\pi \mu_{AB}}} [A][B] = N_A^2 r^2{}_{AB} \sqrt{\frac{8\pi k_B T}{\mu_{AB}}} [A][B] = Z[A][B]$$

From Maxwell Boltzmann distribution it can be deduced that the fraction of collisions with more energy than the activation energy is

$$\frac{-E_q}{ek_B T}.$$

Therefore the rate of a bimolecular reaction for ideal gases will be:

$$r = Z\rho[A][B]\exp\left(\frac{-E_a}{RT}\right)$$

where:

Z is the collision frequency.

ρ is the steric factor, which will be discussed in detail in the next section.

E_a is the activation energy of the reaction.

T is the absolute temperature.

R is gas constant.

The product $Z\rho$ is equivalent to the preexponential factor of the Arrhenius equation.

Validity of the Theory and Steric Factor

Once a theory is formulated, its validity must be tested, that is, compare its predictions with the results of the experiments.

When the expression form of the rate constant is compared with the rate equation for an elementary bimolecular reaction, r=k(T)[A][B], it is noticed that

$$K(T) = N_A^2 \sigma AB \sqrt{\frac{8k_B T}{\pi m_A}} \exp\left(\frac{-E_a}{RT}\right)$$

That expression is similar to the Arrhenius equation, and gives the first theoretical explanation for the Arrhenius equation on a molecular basis. The weak temperature dependence of the preexponential factor is so small compared to the exponential factor that it cannot be measured experimentally, that is, *"it is not feasible to establish, on the basis of temperature studies of the rate constant, whether the predicted $T^{1/2}$ dependence of the preexponential factor is observed experimentally."*

Steric Factor

If the values of the predicted rate constants are compared with the values of known rate constants it is noticed that collision theory fails to estimate the constants correctly and the more complex the molecules are, the more it fails. The reason for this is that particles have been supposed to be spherical and able to react in all directions; that is not true, as the orientation of the collisions is not always the right one. For example in the hydrogenation reaction of ethylene the H_2 molecule must approach the bonding zone between the atoms, and only a few of all the possible collisions fulfill this requirement.

A new concept must be introduced: the steric factor, It is defined as the ratio between the experimental value and the predicted one (or the ratio between the frequency factor and the collision frequency, and it is most often less than unity.

$$\rho = \frac{A_{observed}}{Z_{observed}}$$

Usually, the more complex the reactant molecules, the lower the steric factor. Nevertheless, some reactions exhibit steric factors greater than unity: the harpoon reactions, which involve atoms that exchange electrons, producing ions. The deviation from unity can have different causes: the molecules are not spherical, so different geometries are possible; not all the kinetic energy is delivered into the right spot; the presence of a solvent (when applied to solutions), etc.

The Collision Theory of Reaction Rates

The individual factors which affect the rate of a reaction (temperature, concentration, and so on) are discussed on separate pages. You can get at these via the rates of reaction menu - there is a link at the bottom of the page.

We are going to look in detail at reactions which involve a collision between two species.

Reactions where a single species falls apart in some way are slightly simpler because you won't be involved in worrying about the orientation of collisions. Reactions involving collisions between more than two species are going to be extremely uncommon.

Reactions Involving Collisions Between Two Species

It is pretty obvious that if you have a situation involving two species they can only react together if they come into contact with each other. They first have to collide, and then they *may* react.

Why "*may* react"? It isn't enough for the two species to collide - they have to collide the right way around, and they have to collide with enough energy for bonds to break.

(The chances of all this happening if your reaction needed a collision involving more than 2 particles are remote. All three (or more) particles would have to arrive at exactly the same point in space at the same time, with everything lined up exactly right, and having enough energy to react. That's not likely to happen very often!)

The Orientation of Collision

Consider a simple reaction involving a collision between two molecules - ethene, $CG_2 = CH_2$ and hydrogen chloride, HCl, for example. These react to give chloroethane.

$$CH_2 = CH_2 + HCl \rightarrow CH_3CH_2Cl$$

As a result of the collision between the two molecules, the double bond between the two carbons is converted into a single bond. A hydrogen atom gets attached to one of the carbons and a chlorine atom to the other.

The reaction can only happen if the hydrogen end of the H-Cl bond approaches the carbon-carbon double bond.

Any other collision between the two molecules doesn't work. The two simply bounce off each other.

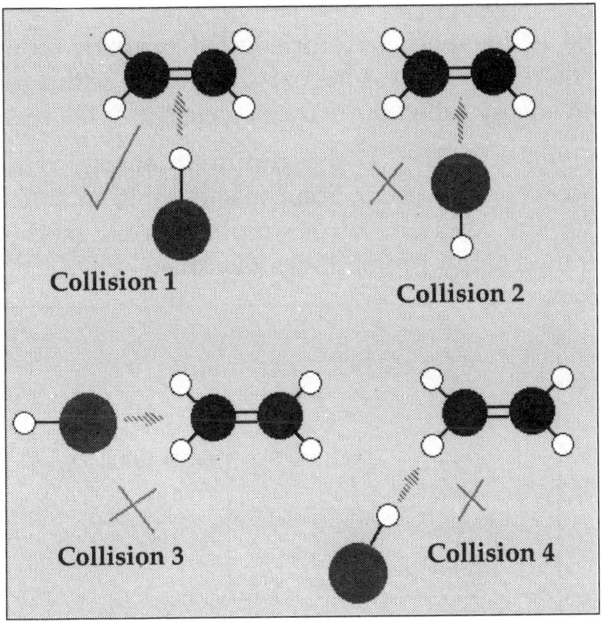

Fig. 10.1

If you haven't read the page about the mechanism of the reaction, you may wonder why collision 2 won't work as well. The double bond has a high concentration of negative charge around it due to the electrons in the bonds. The approaching chlorine atom is also slightly negative because it is more electronegative than hydrogen. The repulsion simply causes the molecules to bounce off each other.

In any collision involving unsymmetrical species, you would expect that the way they hit each other will be important in deciding whether or not a reaction happens.

The Energy of the Collision

Activation Energy

Even if the species are orientated properly, you still won't get a reaction unless the particles collide with a certain minimum energy called the activation energy of the reaction.

Activation energy is the minimum energy required before a reaction can occur. You can show this on an energy profile for the reaction. For a simple over-all exothermic reaction, the energy profile looks like this:

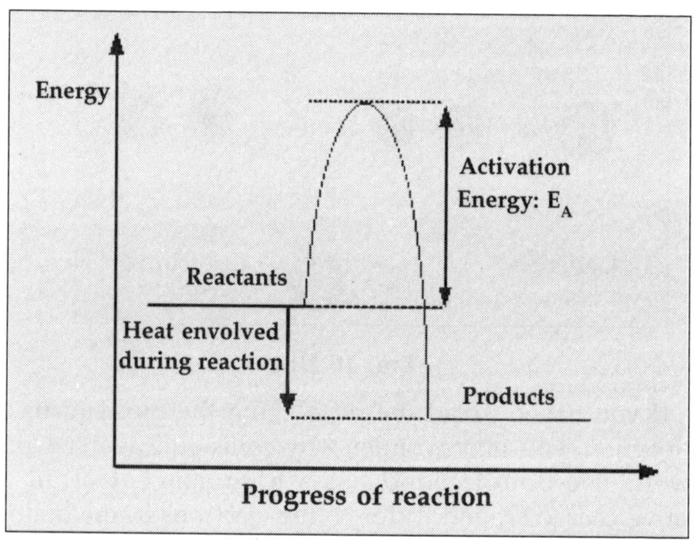

Fig. 10.2

If the particles collide with less energy than the activation energy, nothing important happens. They bounce apart. You can think of the activation energy as a barrier to the reaction. Only those collisions which have energies equal to or greater than the activation energy result in a reaction.

Any chemical reaction results in the breaking of some bonds (needing energy) and the making of new ones (releasing energy). Obviously some bonds have to be broken before new ones can be made. Activation energy is involved in breaking some of the original bonds.

Where collisions are relatively gentle, there isn't enough energy available to start the bond-breaking process, and so the particles don't react.

The Maxwell-Boltzmann Distribution

Because of the key role of activation energy in deciding whether a collision will result in a reaction, it would obviously be useful to know what sort of proportion of the particles present have high enough energies to react when they collide.

In any system, the particles present will have a very wide range of energies. For gases, this can be shown on a graph called the Maxwell-Boltzmann Distribution which is a plot of the number of particles having each particular energy.

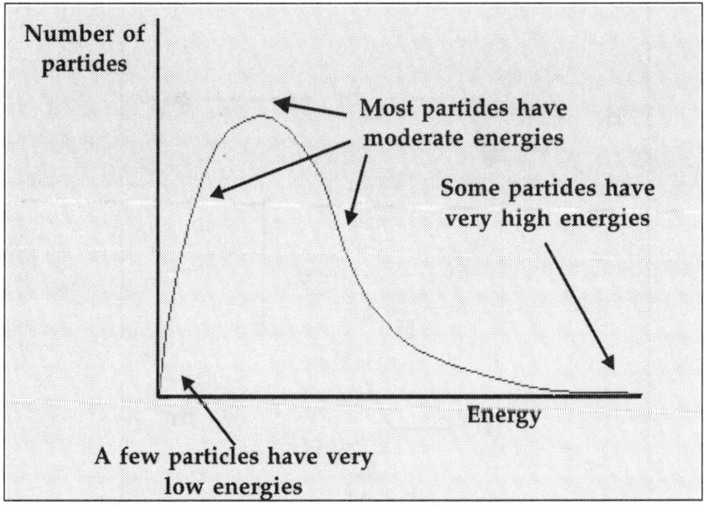

Fig. 10.3

The area under the curve is a measure of the total number of particles present

Transition State

The transition state of a chemical reaction is a particular configuration along the reaction coordinate. It is defined as the state corresponding to the highest energy along this reaction coordinate. At this point, assuming a perfectly irreversible reaction, colliding reactant molecules will always go on to form products The transition state shown below occurs during the SN2 reaction of bromoethane with a hydroxyl anion.

History of Concept

The concept of a transition state has been important in many theories of the rate at which chemical reactions occur. This started with the transition state theory (also referred to as the activated complex theory), which was first developed around 1935 and which introduced basic concepts in chemical kinetics which are still used today.

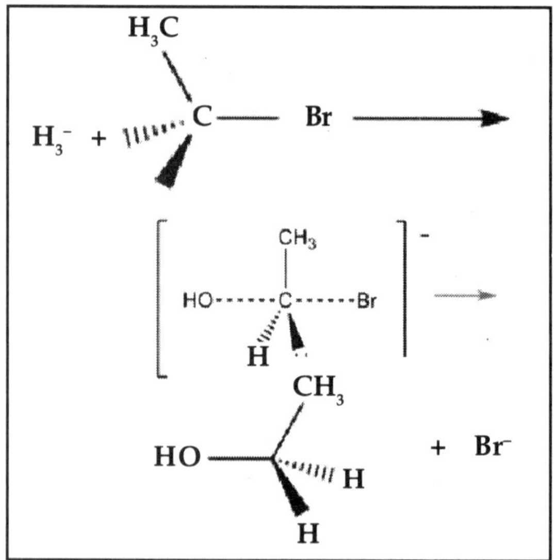

Fig. 10.4

Transition State

Explanation

A collision between reactant molecules may or may not result in a successful reaction. The outcome depends on factors such as the relative kinetic energy, relative orientation and internal energy of the molecules. Even if the collision partners form an activated complex they are not bound to go on and form products, and instead the complex may fall apart back to the reactants.

Observing Transition States

Because of the rules of quantum mechanics, the transition state cannot be captured or directly observed; the population at that point is zero. However, cleverly manipulated spectroscopic techniques can get us as close as the timescale of the technique will allow us. Femtochemical IR spectroscopy was developed for precisely that reason, and it is possible to probe molecular structure extremely close to the transition point. Often along the reaction coordinate reactive intermediates are present not much lower in energy from a transition state making it difficult to distinguish between the two.

Locating Transition States by Computational Chemistry

Transition state structures can be determined by searching for first-order saddle points on the potential energy surface (PES) Such a saddle point is a point where there is a minimum in all dimensions but one. Almost all quantum-chemical methods (DFT, MP2, ...) can be used to find transition states. However, locating them is often difficult and there is no method guaranteed to find the right transition state. There are many different methods of searching for transition states and different quantum chemistry program packages include different ones. Many methods of locating transition states also aim to find the minimum energy pathway (MEP) along the PES. Each method has its advantages and disadvantages depending on the particular reaction under investigation. Summaries of some of the main methods are given below.

Synchronous Transit

There are several type of synchronous transit type methods with the most common being the linear synchronous transit (LST) method and the quadratic synchronous transit (QST). The LST method generates an estimate of the transition state by finding the highest point along shortest line connecting two minima. The QST method extends this further by subsequently searching for a minimum along a line perpendicular to the previous one. The path connecting minima and the found point may then be searched for a saddle point (a maximum).

Nudged Elastic Band

There are many variations on the NEB (nudged elastic band) method including the climbing image nudged elastic band and the elastic band. This method works by guessing the MEP which connects the two stable structures. A discrete number of structures (called images) are placed along the guessed-MEP. These images are moved according to:

(A) the force acting on them perpendicular to the path; and

(B) an artificial spring force keeping the images spaced along the MEP. The highest energy image gives a good estimate of the transition state.

String Method

The string method for locating transition states is similar to the NEB in many ways. It also involves a series of images along a guess of the MEP, but in this case the images are moved in two steps. Firstly, the images are moved according to the force acting on them perpendicular to the path. Using an interpolated path, the images are moved short distances along the MEP to make sure they are evenly space. Variations on the string method include the growing string method in which the guess of the pathway is generated as the programme progresses.

Dimer Method

The dimer method can be used to find possible transition states without knowledge of the final structure or to refine a good guess of a transition structure. The "dimer" is formed by two images very close to each other on the PES. The method works by moving the dimer uphill from the starting position whilst rotating the dimer to find the direction of lowest curvature (ultimately negative).

The Hammond-Leffler Postulate

The Hammond-Leffler Postulate states that the structure of the transition state more closely resembles either the product or the starting material, depending on which is higher in enthalpy.

The Structure-correlation Principle

The structure-correlation principle states that *that structural changes which occur along the reaction coordinate can reveal themselves in the ground state as deviations of bond distances and angles from normal values along the reaction coordinate.* According to this theory if one particular bond length on reaching the transition state increases then this bond is already longer in its ground state compared to a compound not sharing this transition state. One demonstration of this principle is found in the two bicyclic compounds depicted below The one on the left is a bicylco octene which at 200□ C extrudes ethylene in a retro-Diels-Alder reaction.

Compared to the compound on the right (which, lacking an alkene group, is unable to give this reaction) the bridgehead carbon carbon bond length is expected to be shorter if the theory holds because on approaching the transition state this bond gains double bond character. For these two compounds the prediction holds up based on X-ray crystallography and 13C coupling constants (inverse linear relationship with bond length).

Implications for Enzymatic Catalysis

One way in which enzymatic catalysis proceeds is by stabilizing the transition state through electrostatics. By lowering the energy of the transition state, it allows a greater population of the starting material to attain the energy needed to overcome the transition energy and proceed to product.

11

Concentration

In chemistry, concentration is the measure of how much of a given substance there is mixed with another substance. This can apply to any sort of chemical mixture, but most frequently the concept is limited to *homogeneous* solutions, where it refers to the amount of *solute* in the *solvent*.

To concentrate a solution, one must add more solute, or reduce the amount of solvent (for instance, by selective evaporation). By contrast, to dilute a solution, one must add more solvent, or reduce the amount of solute.

Unless two substances are *fully miscible* there exists a concentration at which no further solute will dissolve in a solution. At this point, the solution is said to be saturated. If additional solute is added to a saturated solution, it will not dissolve (except in certain circumstances, when supersaturation may occur). Instead, phase separation will occur, leading to either coexisting phases or a suspension. The point of saturation depends on many variables such as ambient temperature and the precise chemical nature of the solvent and solute.

Analytical concentration includes all the forms of that substance in the solution.

Qualitative Description

Often in informal, non-technical language, concentration is described in a qualitative way, through the use of adjectives such as "dilute" or "weak" for solutions of relatively low concentration and of others like "concentrated" or "strong" for solutions of relatively high concentration. Those terms relate the amount of a substance in a mixture to the observable intensity of effects or properties caused by that substance. For example, a practical rule is that the more concentrated a chromatic solution is, the more intensely colored it is (usually).

Quantitative Notation

For scientific or technical applications, a qualitative account of concentration is almost never sufficient; therefore quantitative measures are needed to describe concentration. There are a number of different ways to quantitatively express concentration; the most common are listed below. They are based on *mass*, *volume*, or *both*. Depending on what they are based on it is not always trivial to convert one measure to the other, because knowledge of the density might be needed to do so. At times this information may not be available, particularly if the temperature varies.

Mass Versus Volume

Some units of concentration—particularly the most popular one, molarity—require knowledge of a substance's volume, which unlike mass is variable depending on ambient temperature and pressure. In fact (partial) molar volume can even be a function of concentration itself. This is why volumes are not necessarily completely additive when two liquids are added and mixed. Volume-based measures for concentration are therefore not to be recommended for non-dilute solutions or problems where relatively large differences in temperature are encountered (e.g. for phase diagrams).

Unless otherwise stated, all the following measurements of volume are assumed to be at a standard state temperature and pressure (for example 25 degrees Celsius at 1 atmosphere or 101.325 kPa). The measurement of mass does not require such restrictions.

Mass can be determined at a precision of < 0.2 mg on a routine basis with an analytical balance and more precise instruments exist. Both solids and liquids are easily quantified by weighing.

The volume of a liquid is usually determined by calibrated glassware such as burettes and volumetric flasks. For very small volumes precision syringes are available. The use of graduated beakers and cylinders is not recommended as their indication of volume is mostly for decorative rather than quantitative purposes. The volume of solids, particularly of powders, is often difficult to measure, which is why mass is the more usual measure. For gases the opposite is true: the volume of a gas can be measured in a gas burette, if care is taken to control the pressure, but the mass is not easy to measure due to buoyancy effects.

Molarity

Molarity (in units of mol/L, molar, or M) or molar concentration denotes the number of moles of a given substance per liter of solution. A capital letter M is used to abbreviate the units of mol/L. For instance:

$$\frac{\text{2.0 moles of dissolved particles}}{\text{4.0 litres of liquid}} = \text{solution of 0.5 mol/L.}$$

The actual formula for molarity is:

$$\frac{\text{Moles of solute}}{\text{Litres of solution}} = \text{Molarity of solution}$$

Such a solution may be described as "0.50 molar." It must be emphasized that a 0.5 molar solution contains 0.5 moles of solute in 1.0 liter of *solution*. This is *not* equivalent to 1.0 liter of solvent. A 0.5 mol/L solution will contain either slightly more or slightly less than 1 liter of solvent because the process of dissolution causes the volume of the liquid to increase or decrease.

Following the SI system of units, the National Institute of Standards and Technology, the United States authority on measurement, considers the term molarity and the unit symbol M to be obsolete, and suggests instead the amount-of-substance concentration (*c*) with units mol/m3 or other units used alongside the SI such as mol/L This recommendation has not been universally implemented in academia or chemistry research yet.

Preparation of a solution of known molarity involves adding an accurately weighed amount of solute to a volumetric flask, adding some solvent to dissolve it, then adding more solvent to fill to the volume mark.

When discussing the molarity of minute concentrations, such as in pharmacological research, molarity is expressed in units of millimolar (mmol/L, mM, 1 thousandth of a molar), micromolar (μmol/L, μM, 1 millionth of a molar) or nanomolar (nmol/L, nM, 1 billionth of a molar).

Although molarity is by far the most commonly used measure of concentration, particularly for dilute aqueous solutions, it does suffer from a number of disadvantages. Masses can be determined with great precision as balances are often very precise. Determining volume is often not as precise. In addition, due to a thermal expansion, the molarity of a solution changes with temperature without adding or removing any mass For non-dilute solutions another problem is that the molar volume of a substance is itself a function of concentration so that volume is not strictly additive.

Molality

Molality (mol/kg, molal, or *m*) denotes the number of moles of *solute* per kilogram of *solvent* (not *solution*). For instance: adding 1.0 mole of solute to 2.0 kilograms of solvent constitutes a solution with a molality of 0.50 mol/kg. Such a solution may be described as "0.50 molal". The term *molal solution* is used as a shorthand for a "one molal solution", i.e. a solution which contains one mole of the solute per 1000 grams of the solvent.

Following the SI system of units, the National Institute of Standards and Technology, the United States authority on measurement, considers the unit symbol *m* to be obsolete, and suggests instead the term 'molality of substance B' (*mB*) with units mol/kg or a related unit of the SI This recommendation has not been universally implemented in academia yet.

Note that molality is sometimes represented by the symbol (*m*), while molarity by the symbol (M). The two symbols are not meant to be confused, and should not be used as symbols for units. The SI unit for molality is mol/kg. (The unit *m* means meter.)

Like other mass-based measures, the determination of molality only requires a good scale, because the masses of both solvent and solute can be obtained by weighing, and molality is independent of the physical conditions like temperature and pressure, providing advantages over molarity.

In a dilute aqueous solution near room temperature and standard atmospheric pressure, the molarity and molality will be very similar in value. This is because 1 kg of water roughly corresponds to a volume of 1 L at these conditions, and because the solution is dilute, the addition of the solute makes a negligible impact on the volume of the solution.

However, in all other conditions, this is usually not the case.

Mole Fraction

The *mole fraction* (also called *molar fraction*) denotes the number of moles of solute as a proportion of the total number of moles in a solution. For instance: 1 mole of solute dissolved in 9 moles of solvent has a mole fraction of 1/10 or 0.1. Mole fractions are *dimensionless* quantities. (The *mole percentage* or *molar percentage*, denoted "mol %" and equal to 100% times the mole fraction, is sometimes quoted instead of the mole fraction.)

This measure is used very frequently in the construction of phase diagrams. It has a number of advantages:

- the measure is not temperature dependent (such as molarity) and does not require knowledge of the densities of the phase(s) involved.
- a mixture of known mole fraction can be prepared by weighing off the appropriate masses of the constituents
- the measure is *symmetrical*: in the mole fractions the roles of 'solvent' and 'solute' are reversed.

As both mole fractions and molality are only based on the masses of the components it is easy to convert between these measures. This is not true for molarity, which requires knowledge of the density.

Mass Percentage (fraction)

Mass percentage denotes the mass of a substance in a mixture as a percentage of the mass of the entire mixture. (Mass fraction x_m can be used instead of mass percentage by dividing mass percentage to 100.) For instance: if a bottle contains 40 grams of ethanol and 60 grams of water, then it contains 40% ethanol by mass or 0.4 mass fraction ethanol. Commercial concentrated aqueous reagents such as acids and bases are often labeled in concentrations of *weight percentage* with the specific gravity also listed. In older texts and references this is sometimes referred to as *weight-weight percentage* (abbreviated as *w/w* or *wt%*). In water pollution chemistry, a common term of measuring total mass percentage of dissolved solids in an aqueous medium is total dissolved solids.

Mass-volume Percentage

Mass-volume percentage, (sometimes referred to as weight-volume percentage or percent weight per volume and often abbreviated as % m/v or % w/v) describes the mass of the solute in g per 100 mL of the resulting solution. Mass-volume percentage is often used for solutions made from a solid solute dissolved in a liquid. For example, a 40% w/v sugar solution contains 40 g of sugar per 100 mL of resulting solution.

Volume-volume Percentage

Volume-volume percentage (sometimes referred to as percent volume per volume and abbreviated as % v/v) describes the volume of the solute in mL per 100 mL of the resulting solution. This is most useful when a liquid - liquid solution is being prepared, although it is used for mixtures of gases as well. For example, a 40% v/v ethanol solution contains 40 mL ethanol per 100 mL total volume. The percentages are only additive in the case of mixtures of ideal gases.

Normality

Normality highlights the chemical nature of salts: in solution, salts dissociate into distinct reactive species (ions such as H+, Fe^{3+}, or Cl^-). Normality accounts for any discrepancy between the concentrations of the various ionic species in a solution. For example, in a salt such as $MgCl^2$, there are two moles of Cl^- for every mole of Mg^{2+}, so the concentration of Cl^- is said to be 2 N (read: "two normal").

Definition

A normal is one gram equivalent of a solute per liter of solution. The definition of a gram equivalent varies depending on the type of chemical reaction that is discussed - it can refer to acids, bases, redox species, and ions that will precipitate.

Usage

It is critical to note that normality measures a single ion which takes part in an overall solute. For example, one could determine the normality of hydroxide or sodium in an aqueous solution of sodium hydroxide, but the normality of sodium hydroxide itself *has no meaning*. Nevertheless it is often used to describe solutions of acids or bases, in those cases it is implied that the normality refers to the H^+ or OH^- ion. For example, 2 Normal sulfuric acid (H_2SO_4), means that the normality of H^+ ions is 2, or that the molarity of the sulfuric acid is 1. Similarly for 1 Molar H_3PO_4 the normality is 3 as it contains three H^+ ions.

Specific Cases

As ions in solution can react through different pathways, there are three common definitions for normality as a measure of reactive species in solution:

- In acid-base chemistry, normality is used to express the concentration of protons or hydroxide ions in the solution. Here, the normality differs from the molarity by an integer value - each solute can produce n equivalents of reactive species when dissolved. For example: 1 M aqueous $Ca(OH)_2$ is 2 N (normal) in hydroxide.
- In redox reactions, normality measures the quantity of oxidizing or reducing agent that can accept or furnish one mole of electrons. Here, the normality scales from the molarity, most commonly, by a fractional value. Calculating the normality of redox species in solution can be challenging.
- In precipitation reactions, normality measures the concentration of ions which will precipitate in a given reaction. Here, the normality scales from the molarity again by an integer value.

Practical Uses

The measure of normality is extremely useful for titrations - given two species that are known to react with a known ratio, one simply needs to scale the volumes of solutions with known normalities to get a complete reaction with the following equation:

$$N_aV_a=N_bV_b$$

However, normality cannot reliably represent an unambiguous measure of the concentration of a solution. Since the measure of normality depends on the reaction that the solute participates in, the same concentration of solute can possess two *different* normalities for two different reactions. For example, Mg^{2+} is 2 N with respect to a Cl^- ion, but it is only 1 N with respect to an O^{2-} ion.

Accordingly, normality is no longer used to represent the concentration of a solution as such. Instead, a solution should be labeled according to its molarity, and it is then possible to calculate the normality for a particular titration using the equation above. NIST has also stipulated that this unit is obsolete and recommends discontinuing its use.

Equivalents

Expression of concentration in equivalents per liter (or more commonly, milliequivalents per liter) is based on the same principle as normality. A normal solution is one equivalent per liter of solution (Eq/L). The use of equivalents and milliequivalents as a means of expressing concentration is losing favor, but medical reporting of serum concentrations in mEq/L still occurs.

Formal

The *formal* (F) is yet another measure of concentration similar to molarity. Formal concentrations are sometimes used when solving chemical equilibrium problems. It is calculated based on the formula weights of chemicals per liter of solution. The difference between formal and molar concentrations is that the formal concentration indicates moles of the original chemical formula in solution, without regard for the species that actually exist in solution. Molar concentration, on the other hand, is the concentration of species in solution.

For example: if one dissolves sodium carbonate (Na_2CO_3) in a litre of water, the compound dissociates into the Na^+ and CO_3^{2-} ions. Some of the CO_3^{2-} reacts with the water to form HCO_3^- and H_2CO_3. If the pH of the solution is low, there is practically no Na_2CO_3 left in the solution. So, although we have added 1 mol of Na_2CO_3 to the solution, it does not contain 1 M of that substance. (Rather, it contains a molarity based on the other constituents of the solution.) However, it was once said that such solutions contain 1 F of Na_2CO_3.

"Parts-per" Notation

The parts-per notation is used in some areas of science and engineering because it does not require conversion from weights or volumes to more chemically relevant units such as normality or molarity. It describes the amount of one substance in another. It is the ratio of the amount of the substance of interest to the amount of that substance plus the amount of the substance it is in.

- *Parts per hundred* (denoted by '%' [the per cent symbol], and very rarely 'pph') - denotes the amount of a given substance in a total amount of 100 regardless of the units of measure as long as they are the same. e.g. 1 gram per 100 gram. 1 part in 10^2.
- *Parts per thousand* (denoted by '‰' [the per mille symbol], and occasionally 'ppt', though this should be avoided) denotes the amount of a given substance in a total amount of 1000 regardless of the units of measure as long as they are the same. e.g. 1 milligram per gram, or 1 gram per kilogram. 1 part in 10^3.

- *Parts per million* ('ppm') denotes the amount of a given substance in a total amount of 1,000,000 regardless of the units of measure used as long as they are the same. e.g. 1 milligram per kilogram. 1 part in 10^6.
- *Parts per billion* ('ppb') denotes the amount of a given substance in a total amount of 1,000,000,000 regardless of the units of measure as long as they are the same. e.g. 1 milligram per tonne. 1 part in 10^9.
- *Parts per trillion* ('ppt') denotes the amount of a given substance in a total amount of 1,000,000,000,000 regardless of the units of measure as long as they are he same. e.g. 1 milligram per kilotonne. 1 part in 10^{12}.
- *Parts per quadrillion* ('ppq') denotes the amount of a given substance in a total amount of 1,000,00,000,000,000 regardless of the units of measure as ong as they are the same. e.g. 1 milligram per megatonne. 1 part in 10^{15}.

Notes for clarityThe notation is used for convenience and the units of measure must be denoted for clarity though this is frequently not the case even in technical publications.In atmospheric chemistry and in air pollution regulations, the parts per notation is commonly expressed with a v following, such as ppmv, to indicate parts per million by volume. This works finc for gas concentrations (e.g., ppmv of carbon dioxide in the ambient air) but, for concentrations of non-gaseous substances such as aerosols, cloud droplets, and particulate matter in the ambient air, the concentrations are commonly expressed as $\mu g/m^3$ or mg/m^3 (e.g., µg or mg of particulates per cubic metre of ambient air). This expression eliminates the nccd to take into account the impact of temperature and pressure on the density and hence weight of the gas.

The usage is generally qite fixed inside most specific branches of science, leading some researchers to believe that their own usage (mass/mass, volume/volume or others) is the only correct one. This, in turn, leads them not to specify their usage in their research, and others may therefore misinterpret their results. For example, electrochemists often use volume/volume, while chemical engineers may use mass/ mass as well as volume/volume. Many academic papers of otherwise excellent level fail to specify their usage of the part-per notation. The difference between expressing concentrations as mass/mass or volume/volume is quite significant when dealing with gases and it is very important to specify which is being used. It is quite simple, for example, to distinguish ppm by volume from ppm by mass or weight by using *ppmv* or *ppmw*.

12

Category
Reagents for Organic Chemistry

This category was created to provide a "home" for inorganic compounds (such as $NaBH_4$) that are widely used in stoichiometric quantities in organic chemistry, but widely used organic reagents (such as oxalyl chloride) may belong here also. This category is not for catalysts such as Pd.

Subcategories

This category has the following 9 subcategories, out of 9 total.

A Alkylating agents

D Dehydrating agents

E Ethylating agents

F Fluorinating agents

M Methylating agents

O Organomagnesium compounds

	Oxidizing agents
P	Protecting groups
R	Reducing agents

Category "Reagents for Organic Chemistry"

A

AD-mix

Acetyl chloride

Aluminium chloride

Armstrong's acid

Azobisisobutyronitrile

B

9-Borabicyclononane

BOP reagent

Benzyl chloroformate

Benzyltrimethylammonium fluoride

1,8-Bis (dimethylamino) naphthalene

(Bis (trifluoroacetoxy) iodo) benzene

Bis (trimethylsilyl) acetamide

Bis (trimethylsilyl) amine

N-Bromosuccinimide

Burgess reagent

Tert-Butyl hydroperoxide

N-Butyllithium

Tert-Butyllithium

C

CBS catalyst

Caesium acetate

Carbon tetraiodide

Carbonyldiimidazole

Chloromethyl methyl ether

Chlorosulfonyl isocyanate

Chlorosulfuric acid

Copper chromite

Copper(I)thiophene-2-carboxylate

Cornforth reagent

Cyclohexenone

D

Danishefsky's diene

Dansyl chloride

Di-tert-butyl dicarbonate

1, 5-Diazabicyclo (4.3.0) non-5-ene

1, 8-Diazabicycloundec-7-ene

Dibutylboron triflate

Dichlorocarbene

Diethyl azodicarboxylate

Diethylaminosulfur trifluoride

Diethylzinc

Diisopinocampheylborane

N, N'-Diisopropylcarbodiimide

2, 2-Dimethoxypropane
Dimethyl malonate
4-Dimethylaminopyridine
Dimethyldioxirane
2, 4-Dinitrophenylhydrazine
Diphenylphosphoryl azide
2, 2'-Dipyridyldisulfide
Disiamylborane

E

Edman degradation
Eschenmoser's salt
1, 2-Ethanedithiol
Ethyl bromoacetate
Ethyl chloroformate

F

Fluorenylmethyloxy-carbonyl chloride
Frémy's salt

G

Gilman reagent
Grignard reaction
Grignard reagents

H

Hexamine
Hydroxybenzotriazole
Hypophosphorous acid

I

2-Iodoxybenzoic acid

L

Lawesson's reagent
Lithium bis(trimethylsilyl)amide
Lithium diisopropylamide

M

Methanesulfonic anhydride
Methanesulfonyl chloride
Methoxymethylene-triphenylphosphine
Methyl methanesulfonate
N-Methylmorpholine N-oxide
N-Methylmorpholine

N

N-Chlorosuccinimide
Naphthalen-1, 8-diyl 1, 3, 2, 4-dithiadiphosphetane 2, 4-disulfide
Ninhydrin
Nysted reagent

O

Organolithium reagent

P

Pempidine
Petasis reagent
Phosgene
Phosphorus tribromide
Polymethylhydrosiloxane
P cont.
Potassium tert-butoxide
1, 3-Propanedithiol
PyBOP

R

Reagent

Reinecke's salt

Rieke metals

S

Selectfluor

Seliwanoff's test

Sodium bisulfite

Sodium hydride

Sodium methylsulfinylmethylide

Sulfolene

T

TASF reagent

Tebbe's reagent

Tetra-n-butylammonium fluoride

Tetrakis(tripheny-lphosphine)- palladium(0)

1, 1, 3, 3-Tetramethyl-guanidine

2, 2, 6, 6-Tetramethyl-piperidine

Thioacetic acid

Thionyl chloride

Titanium tetrachloride

P-Toluenesulfonic acid

4-Toluenesulfonyl chloride

Tribromoisocyanuric acid

2, 2, 2-Trichlorethoxy-carbonyl chloride

Triethyl phosphonoacetate

Triethyloxonium tetrafluoroborate

Triflic anhydride

Trifluoro-methanesulfonic acid

Trimethylsilyl azide

Trimethylsilyldiazomethane

Triphenyltin hydride

Tris-(benzyltriazoly-lmethyl) amine

Trost ligand

V

Vanadium tetrachloride

W

Webster's reagent

Wollins' reagent

Limiting Reagent

In chemistry, the limiting reagent, also known as the "limiting reactant", is the chemical that determines how far the reaction will go before the chemical in question gets "used up", causing the reaction to stop. The chemical of which there are fewer moles than the proportion requires is the limiting reagent.

Example

Consider the combustion of benzene:

$2C_6H_6 + 15O_2 \quad 12CO_2 + 6H_2O$

This means that 15 mol O_2 is required to react with 2 mol C_6H_6.

The amount of oxygen required for other quantities of benzene can be calculated using cross-multiplication (the rule of three). For example, if 1.5 mol C_6H_6 is present, 11.25 mol O_2 is required:

$$1.5 \text{ mol } C_6H_6 \times \frac{15 mol O_2}{2 mol C_6H_6} = 11.25 mol O_2$$

If only 7 mol O_2 is present, the oxygen will be consumed before benzene. Therefore, O_2 must be the limiting reagent.

This conclusion can be verified by comparing the mole ratio of O_2 and C_6H_6 required by the balanced equation with the mole ratio actually present:

$$\text{required: } \frac{mol O_2}{mol C_6H_6} = \frac{15 mol O_2}{2 mol C_6H_6} 7.5 mol O_2$$

$$\text{actual: } \frac{mol O_2}{mol C_6H_6} = \frac{15 mol O_2}{2 mol C_6H_6} 11.25 mol O_2$$

Since the actual ratio is too small, O_2 is the limiting reagent.

Consider a Typical Thermite Reaction

If 20.0 g of Fe_2O_3 are reacted with 8.00 g Al(s) in the thermite reaction, Which reactant is limiting?.

$Fe_2O_3(s) + 2Al(s) \rightarrow 2\ Fe\ (l) + Al_2O_3\ (s)$

First, determine how many moles of Fe(l) can be produced from either reactant.

Moles produced of Fe from reactant Fe_2O_3

$$mol\ Fe_2O_3 = \frac{gramsFe_2O_3}{g/molFe_2O_3}$$

$$mol\ Fe_2O_3 = \frac{20.0g}{159.7g/mol} = 0.125mol$$

$$mol\ Fe = (2)\ (0.125) = 0.250molFe$$

Moles produced of Fe from reactant Al

$$\text{mol Al} = \frac{gramsAl}{g/molAl}$$

$$\text{mol Al} = \frac{8.00g}{26.98g/mol} = 0.297mol$$

$$\text{mol Fe} = \frac{(2)(0.297)}{2} = 0.297molFe$$

Because the moles Fe produced from $Fe_2O_3(0.254\ mol)$ is less than the moles Fe produced from $Al(0.297mol)$, Fe_2O_3 is the limiting reagent.

By looking at chemical equation for the thermite reaction, the limiting reagent can be found based on the ratio of moles of one reactant to another and the total atomic mass of the reactant compounds.

Limiting Reagent Formula

There is a much simpler formula which can be used. However, you must first calculate the moles of both of the reagents in the reactio. Once the number of moles have been figured out, just simplyfill in this equation (reagent 1 being the first reactant and 2 being the second):

$$\text{Moles of Reagent 2} \times \frac{\text{Coefficient of Reagent 1}}{\text{Coefficient of Reagent 2}} - \text{Moles of Reagent 2}$$

When the answer to the formula is less than zero, reagent 1 is the excss reagent. When the answer is larger than zero, reagent 1 is thelimiting reagent. The number shows how much in excess one reagent is from another. If the answer for the formula is zero, both reagents are perfectly balanced. The unit of an answer is in *moles*.

13

Solvent

A solvent is a liquid or gas that dissolves a solid, liquid, or gaseous solute, resulting in a solution.

The most common solvent in everyday life is water. Most other commonly-used solvents are organic (carbon-containing) chemicals. These are called organic solvents. Solvents usually have a low boiling point and evaporate easily or can be removed by distillation, leaving the dissolved substance behind. To distinguish between solutes and solvents, solvents are usually present in the greater amount. Solvents can also be used to extract soluble compounds from a mixture, the most common example is the brewing of coffee or tea with hot water. Solvents are usually clear and colorless liquids and many have a characteristic odor. The concentration of a solution is the amount of compound that is dissolved in a certain volume of solvent. The solubility is the maximal amount of compound that is soluble in a certain volume of solvent at a specified temperature. Common uses for organic solvents are in dry cleaning (e.g. tetrachloroethylene), as paint thinners (e.g. toluene, turpentine), as nail polish removers and glue solvents (acetone, methyl acetate, ethyl acetate), in spot removers (e.g. hexane, petrol ether), in detergents (citrus terpenes), in perfumes (ethanol), and in chemical syntheses. The use of inorganic solvents (other than water) is typically limited to research chemistry and some technological processes.

In 2005, the worldwide market for solvents had a total volume of around 17.9 million tons, which led to a turnover of about 8 billion Euro.

Solutions and Solvation

When one substance is dissolved into another, a solution is formed. This is opposed to a mixture where one compound is added to another and no chemical bond is formed; a way to think of mixtures and solutions is to compare a cup of water with sand mixed in versus a soda where all of the ingredients are uniform to create a new substance. No residue is left in the bottom. The mixing is referred to as miscibility, whereas the ability to dissolve one compound into another is known as solubility. However, in addition to mixing, both substances in the solution can interact with each other in specific ways. Solvation describes these interactions. When something is dissolved, molecules of the solvent arrange themselves around molecules of the solute. Heat is evolved and entropy is increased making the solution more thermodynamically stable than the solute alone. This arranging is mediated by the respective chemical properties of the solvent and solute, such as hydrogen bonding, dipole moment and polarizability.

Solvent Classifications

Solvents can be broadly classified into two categories: *polar* and *non-polar*. Generally, the dielectric constant of the solvent provides a rough measure of a solvent's polarity. Solvents with a dielectric constant of less than 15 are generally considered nonpolar. Technically, the dielectric constant measures the solvent's ability to reduce the field strength of the electric field surrounding a charged particle immersed in it. This reduction is then compared to the field strength of the charged particle in a vacuum In laymen's terms, dielectric constant of a solvent can be thought of as its ability to reduce the solute's internal charge.

Other Polarity Scales

Dielectric constants are not the only measure of polarity. Because solvents are used by chemists to carry out chemical reactions or observe chemical and biological phenomena, more specific measures of polarity are required.

- *The Grunwald Winstein m***Y** *scale* measures polarity in terms of solvent influence on buildup of positive charge of a solute during a chemical reaction.
- *Kosower's* **Z** *scale* measures polarity in terms of the influence of the solvent on uv absorption maxima of a salt, usually pyridinium iodide or the pyridinium zwitterion.
- *Donor number and donar acceptor scale* measures polarity in terms of how a solvent interacts with specific substances, like a strong Lewis acid or a strong Lewis base.

The polarity, dipole moment, polarizability and hydrogen bonding of a solvent determines what type of compounds it is able to dissolve and with what other solvents or liquid compounds it is miscible. As a rule of thumb, polar solvents dissolve polar compounds best and non-polar solvents dissolve non-polar compounds best: "like dissolves like". Strongly polar compounds like sugars (e.g. sucrose) or ionic compounds, like inorganic salts (e.g. table salt) dissolve only in very polar solvents like water, while strongly non-polar compounds like oils or waxes dissolve only in very non-polar organic solvents like hexane. Similarly, water and hexane (or vinegar and vegetable oil) are not miscible with each other and will quickly separate into two layers even after being shaken well.

Polar Protic and Polar Aprotic

Solvents with a relative static permittivity greater than 15 can be further divided into protic and aprotic. Protic solvents solvate anions (negatively charged solutes) strongly via hydrogen bonding. Water is a protic solvent. Aprotic solvents such as acetone or dichloromethane tend to have large dipole moments (separation of partial positive and partial negative charges within the same molecule) and solvate positively charged species via their negative dipole In chemical reactions the use of polar protic solvents favors the S_N1 reaction mechanism, while polar aprotic solvents favor the S_N2 reaction mechanism.

Solvent Effects

Boiling Point

Another important property of solvents is boiling point. This also determines the speed of evaporation. Small amounts of low-boiling solvents like diethyl ether, dichloromethane, or acetone will evaporate in seconds at room temperature, while high-boiling solvents like water or dimethyl sulfoxide need higher temperatures, an air flow, or the application of vacuum for fast evaporation.

- Low Boilers: Boiling ranges below 100 °C.
- Medium Boilers: Boiling ranges between 100 °C and 150 °C.
- High Boilers: Boiling ranges above 150 °C.

For comparison, the boiling point of water is 100 °C.

Density

Most organic solvents have a lower density than water, which means they are lighter and will form a separate layer on top of water. An important exception: many halogenated solvents like dichloromethane or chloroform will sink to the bottom of a container, leaving water as the top layer. This is important to remember when partitioning compounds between solvents and water in a separatory funnel during chemical syntheses.

Health and Safety

Fire

Most organic solvents are flammable or highly flammable, depending on their volatility. Exceptions are some chlorinated solvents like dichloromethane and chloroform. Mixtures of solvent vapors and air can explode. Solvent vapors are heavier than air; they will sink to the bottom and can travel large distances nearly undiluted. Solvent vapours can also be found in supposedly empty drums and cans, posing a flash fire hazard; hence empty containers of volatile solvents should be stored open and upside down.

Both diethyl ether and carbon disulfide have exceptionally low autoignition temperatures which increase greatly the fire risk associated with these solvents. The autoignition temperature of carbon disulfide is below 100°C (212°F), so as a result objects such as steam pipes, light bulbs, hotplates and recently extinguished bunsen burners are able to ignite its vapours.

Peroxide Formation

Ethers like diethyl ether and tetrahydrofuran (THF) can form highly explosive organic peroxides upon exposure to oxygen and light, THF is normally more able to form such peroxides than diethyl ether. One of the most susceptible solvents is diisopropyl ether.

The heteroatom (oxygen) stabilizes the formation of a free radical which is formed by the abstraction of a hydrogen atom by another free radical. The carbon centred free radical thus formed is able to react with an oxygen molecule to form a peroxide compound. A range of tests can be used to detect the presence of a peroxide in an ether; one is to use a combination of iron sulfate and potassium thiocyanate. The peroxide is able to oxidize the Fe^{2+} ion to an Fe^{3+} ion which then form a deep red coordination complex with the thiocyanate. In extreme cases the peroxides can form crystalline solids within the vessel of the ether.

Unless the desiccant used can destroy the peroxides, they will concentrate during distillation due to their higher boiling point. When sufficient peroxides have formed, they can form a crystalline and shock sensitive solid precipitate. When this solid is formed at the mouth of the bottle, turning the cap may provide sufficient energy for the peroxide to detonate. Peroxide formation is not a significant problem when solvents are used up quickly; they are more of a problem for laboratories which take years to finish a single bottle. Ethers have to be stored in the dark in closed canisters in the presence of stabilizers like butylated hydroxytoluene (BHT) or over sodium hydroxide.

Peroxides may be removed by washing with acidic iron(II) sulfate, filtering through alumina, or distilling from sodium/benzophenone. Alumina does not destroy the peroxides; it merely traps them. The advantage of using sodium/benzophenone is that moisture and oxygen is removed as well.

Health Effects

Many solvents can lead to a sudden loss of consciousness if inhaled in large amounts. Solvents like diethyl ether and chloroform have been used in medicine as anesthetics, sedatives, and hypnotics for a long time. Ethanol (grain alcohol) is a widely used and abused psychoactive drug. Diethyl ether, chloroform, and many other solvents (e.g. from gasoline or glues) are used recreationally in glue sniffing, often with harmful long term health effects like neurotoxicity or cancer. Methanol can cause internal damage to the eyes, including permanent blindness.

It is interesting to note that ethanol has a synergistic effect when taken in combination with many solvents. For instance a combination of toluene/benzene and ethanol causes greater nausea/vomiting than either substance alone. Many chemists make a point of not drinking beer/wine/other alcoholic drinks if they know that they have been exposed to an aromatic solvent.

Environmental Contamination

A major pathway to induce health effects arises from spills or leaks of solvents that reach the underlying soil. Since solvents readily migrate substantial distances, the creation of widespread soil contamination is not uncommon; there may be about 5000 sites worldwide that have major subsurface solvent contamination; this is particularly a health risk if aquifers are affected.

Chronic Health Effects

Some solvents including chloroform and benzene (an ingredient of gasoline) are carcinogenic. Many others can damage internal organs like the liver, the kidneys, or the brain.

General Precautions

- Avoid being exposed to solvent vapors by working in a fume hood, or with local exhaust ventilation (LEV), or in a well ventilated area.
- Keep the storage containers tightly closed.
- Never use open flames near flammable solvents; use electrical heating instead.
- Never flush flammable solvents down the drain; read safety data sheets for proper disposal information.
- Avoid the inhalation of solvent vapors.
- Avoid contact of the solvent with the skin — many solvents are easily absorbed through the skin. They also tend to dry the skin and may cause sores and wounds.

Solution

In chemistry, a solution is a homogeneous mixture composed of two or more substances. In such a mixture, a solute is dissolved in another substance, known as a solvent. A common example is a solid, such as salt or sugar, dissolved in water, a liquid. Gases may dissolve in liquids, for example, carbon dioxide or oxygen in water. Liquids may dissolve in other liquids. Gases can combine with other gases to form mixtures, rather than solutions. All solutions are characterized by interactions between the solvent phase and solute molecules or ions that result in a net decrease in free energy. Under such a definition, gases typically cannot function as solvents, since in the gas phase interactions between molecules are minimal due to the large distances between the molecules. This lack of interaction is the reason gases can expand freely and the presence of these interactions is the reason liquids do not expand.

Examples of solid solutions are alloys and certain minerals and polymers containing plasticizers. The ability of one compound to dissolve in another compound is called solubility. The physical properties of compounds such as melting point and boiling point change when other compounds are added. Together they are called colligative properties. There are several ways to quantify the amount of one compound dissolved in the other compounds collectively called concentration. Examples include *molarity, molality,* mole fraction, and *parts per million* (ppm).

Solutions should be distinguished from non-homogeneous mixtures such as colloids and suspensions. When a liquid is able to completely dissolve in another liquid the two liquids are miscible. Two substances that can never mix to form a solution are called immiscible.

Solvation

During solvation, especially when the solvent is polar, a structure forms around it, which allows the solute-solvent interaction to remain stable.

When no more of a solute can be dissolved into a solvent, the solution is said to be saturated. However, the point at which a solution can become saturated can change significantly with different environmental factors, such as temperature, pressure, and contamination. For some solute-solvent combinations a supersaturated solution can be prepared by raising the solubility (for example by increasing the temperature) to dissolve more solute, and then lowering it (for example by cooling).

Usually, the greater the temperature of the solvent, the more of a given solid solute it can dissolve. However, most gases and some compounds exhibit solubility that decrease with increased temperature. Such behavior is a result of an exothermic enthalpy of solution. Some surfactants exhibit this behaviour. The solubility of liquids in liquids is generally less temperature-sensitive than that of solids or gases.

Ideal Solutions

Properties of an ideal solution can be calculated by the linear combination of the properties of its components.

If both solute and solvent exist in equal quantities (such as in a 50% ethanol, 50% water solution), the concepts of "solute" and "solvent" become less relevant, but the substance that is more often used as a solvent is normally designated as the solvent (in this example, water).

Mixture

In chemistry, a mixture is when two or more different substances are mixed together but not combined chemically. The molecules of two or more different substances are mixed in the from of solutions, suspensions, and colloids.

While there are no physical changes in a mixture, the chemical properties of a mixture, such as its melting point, may differ from those of its components. Mixtures can usually be separated into its original components by mechanical means. Mixtures are either homogeneous or heterogeneous.

Mixtures are the product of a blending or mixing of chemical substances like elements and compounds, without chemical bonding or other chemical change, so that each ingredient substance retains its own chemical properties and makeup.

Suspensions

A heterogeneous mixture is where it is not evenly distributed within the mixture. A suspension is when the particles of one substance are suspended in the other substance (the two substances do not mix into a 'seam-less' mixture- a 'whole'). An example of a suspension is putting flour in water. You can see the water as a separate substance from the particles of flour (the flour is obviously not blended with the water). One Example Is Salad. If you mix the salad it doesn't hold a form but still mixes. Another example is a cake.

Colloidal Dispersions

Colloids are homogeneous mixtures in which the particles of one or more components have at least one dimension in the range of 1 to 1000 nm, larger than those in a solution but smaller than those in a suspension. Colloids are the same as suspensions, except they don't leave sediments. In general, a colloid or colloidal dispersion is a substance with components of one or two phases. It creates the Tyndall effect when light passes through it. A colloid will not settle. Jelly, milk, blood, paint, fog, shampoo, and glue are examples of colloid dispersions.

Mixtures are made of two or more substances - elements, compounds, or both - that are together in the same place but are not chemically combined. Mixtures differ from compounds in two ways. Elements and compounds are pure subsances but most of the materials you see every day are not. Instead they are mixtures.

Mixtures and Compounds

A compound is not a mixture. A compound has very different properties than the elements it is made of, but a mixture contains several substances which keep their properties.

14

Chemical Potential

In thermodynamics, physics and chemistry, chemical potential, symbolized by μ, is a term introduced by the American engineer, chemist and mathematical physicist Josiah Williard Gibbs, which he defined as follows:

> "If to any homogeneous mass in a state of hydrostatic stress we suppose an infinitesimal quantity of any substance to be added, the mass remaining homogeneous and its entropy and volume remaining unchanged, the increase of the energy of the mass divided by the quantity of the substance added is the *potential* for that substance in the mass considered".

Gibbs noted also that for the purposes of this definition, any chemical element or combination of elements in given proportions may be considered a substance, whether capable or not of existing by itself as a homogeneous body. Chemical potential is also referred to as partial molar gibbs energy.

In modern statistical physics the chemical potential is the lagrange multiplier for the average particle constraint, when maximizing, the entropy. This is the precise and abstract scientific definition, just like the temperature is defined as the lagrange multiplier for the average energy constraint.

Example

Various thermodynamic properties determine what the chemical potential is. For example, consider charged particles in a fluid. A concentration gradient in a fluid may promote movement of particles in one direction, and the electric potential gradient may promote movement of the particles in another. The chemical potential would account for both concentration and electric components and describe a potential distribution that determines net particle movement.

History

In his 1873 paper A Method of Geometrical Representation of the Thermodynamic Properties of Substances by Means of Surfaces Gibbs introduced the preliminary outline of the principles of his new equation able to predict or estimate the tendencies of various natural processes to ensue when bodies or systems are brought into contact. By studying the interactions of homogeneous substances in contact, i.e. bodies, being in composition part solid, part liquid, and part vapor, and by using a three-dimensional volume–entropy–internal energy graph, Gibbs was able to determine three states of equilibrium, i.e. "necessarily stable", "neutral", and "unstable", and whether or not changes will ensue. In 1876, Gibbs built on this framework by introducing the concept of chemical potential so to take into account chemical reactions and states of bodies which are chemically different from each other. In his own words, to summarize his results in 1873, Gibbs states:

If we wish to express in a single equation the necessary and sufficient condition of thermodynamic equilibrium for a substance when surrounded by a medium of constant pressure *P* and temperature *T*, this equation may be written:

$$\delta(\varepsilon - T\eta + Pv) = 0$$

when δ refers to the variation produced by any variations in the state of the parts of the body, and (when different parts of the body are in different states) in the proportion in which the body is divided between the different states. The condition of stable equilibrium is that the value of the expression in the parenthesis shall be a minimum.

In this description, as used by Gibbs, ε refers to the internal energy of the body, η refers to the entropy of the body, and v is the volume of the body.

Related Terms

The precise meaning of the term *chemical potential* depends on the context in which it is used.

- When speaking of thermodynamic systems, *chemical potential* refers to the *thermodynamic chemical potential*. In this context, the chemical potential is the change in a characteristic thermodynamic state function per change in the number of molecules. Depending on the experimental conditions, the characteristic thermodynamic state function is either: *internal energy, enthalpy, Gibbs energy*, or *Helmholtz energy*. This particular usage is most widely used by experimental chemists, physicists, and chemical engineers.
- Theoretical chemists and physicists often use the term *chemical potential* in reference to the *electronic chemical potential*, which is related to the functional derivative of the *density functional*, sometimes called the *energy functional*, found in Density Functional Theory. This particular usage of the term is widely used in the field of *electronic structure theory*.
- Physicists sometimes use the term *chemical potential* in the description of relativistic systems of fundamental particles.

Thermodynamic Chemical Potential

The chemical potential of a thermodynamic system is the amount by which the energy of the system would change if an additional particle were introduced, with the entropy and volume held fixed. If a system contains more than one species of particle, there is a separate chemical potential associated with each species, defined as the change in energy when the number of particles *of that species* is increased by one. The chemical potential is a fundamental parameter in thermodynamics and it is conjugate to the particle number.

The chemical potential is particularly important when studying systems of reacting particles. Consider the simplest case of two species, where a particle of species 1 can transform into a particle of species 2 and vice versa. An example of such a system is a saturated mixture of water liquid (species 1) and water vapor (species 2). If the system is at equilibrium, the chemical potentials of the two species must be equal. Otherwise, a net release of energy in the form of heat would occur when the species of higher potential transforms into the other species, and a net gain of energy (again in the form of heat) would occur for the reverse transformation. In chemical reactions, the equilibrium conditions are generally more complicated because more than two species are involved.

In this case, the relation between the chemical potentials at equilibrium is given by the law of mass action.

Since the chemical potential is a thermodynamic quantity, it is defined independently of the microscopic behavior of the system, i.e. the properties of the constituent particles. However, some systems contain important variables that are equivalent to the chemical potential. In Fermi gases and Fermi liquids, the chemical potential at zero temperature is equivalent to the Fermi energy. In electronic systems, the chemical potential is related to an effective electrical potential.

A way to understand the chemical potential is to consider one mole of methane and 2 moles of oxygen. If a flame is brought near this mixture, the following reaction will occur:

$$CH_4 + 2\,O_2 \rightarrow CO_2 + 2\,H_2O$$

and energy (heat) will be released. This energy comes from the difference in chemical potential between CH_4 and O_2 on one hand (higher potential) and CO_2 and H_2O on the other hand (lower). The whole energy that will be released will be given by

$$\mu(CH_4) + 2\,\mu(O_2) - \mu(CO_2) - 2\,\mu(H_2O)$$

Similar examples can be found within batteries where chemical energy is converted into electrical energy.

Precise Definition

Consider a thermodynamic system containing n constituent species. Its total internal energy U is postulated to be a function of the entropy S, the volume V, and the number of particles of each species $N1, \ldots, Nn$

$$U = U(S, V, N_1, \ldots Nn)$$

By referring to U as the *internal energy*, it is emphasized that the energy contributions resulting from the interactions between the system and external objects are excluded. For example, the gravitational potential energy of the system with the Earth are not included in U.

The chemical potential of the i-th species, μi is defined as the partial derivative

$$\mu_i = \left(\frac{\partial U}{\partial N_i}\right)_{S,V,N_j \neq i}$$

where the subscripts simply emphasize that the entropy, volume, and the other particle numbers are to be kept constant.

In real systems, it is usually difficult to hold the entropy fixed, since this involves good thermal insulation. It is therefore more convenient to define the Helmholtz energy *A*, which is a function of the temperature *T*, volume, and particle numbers:

$$A = A(T, V, N_1, \dots Nn)$$

In terms of the Helmholtz energy, the chemical potential is

$$\mu_i = \left(\frac{\partial A}{\partial N_i}\right)_{T,V,N_j \neq i}$$

Laboratory experiments are often performed under conditions of constant temperature and pressure. Under these conditions, the chemical potential is the partial derivative of the Gibbs energy with respect to number of particles

$$\mu_i = \left(\frac{\partial G}{\partial N_i}\right)_{T,p,N_j \neq i}$$

A similar expression for the chemical potential can be written in terms of partial derivative of the enthalpy (under conditions of constant entropy and pressure).

Here, the chemical potential has been defined as the «energy» per molecule. A variant of this definition is to define the chemical potential as the «energy» per mole.

Electronic Chemical Potential

The electronic chemical potential is the functional derivative of the density functional with respect to the electron density.

$$\mu(r) = \left(\frac{\partial E[p]}{\partial p(r)} \right)_{p=p_{ref}}$$

Formally, a functional derivative yields many functions, but is a particular function when evaluated about a reference electron density - just as a derivate yields a function, but is a particular number when evaluated about a reference point. The density functional is written as

$$E[p] = \int p(r)v(r)d^3r + F[p]$$

where $v(r)$ is the *external potential*, e.g., the electrostatic potential of the nuclei and applied fields, and F is the *Universal functional*, which describes the electron–electron interactions, e.g., electron Coulomb repulsion, kinetic energy, and the non-classical effects of exchange and correlation. With this general definition of the density functional, the chemical potential is written as

$$\mu(r) = v(r) + \left[\frac{\partial F[p]}{\partial p(r)} \right]_{p=p_{ref}}$$

Thus, the electronic chemical potential is the effective electrostatic potential experienced by the electron density.

The ground state electron density is determined by a *constrained* variational optimization of the electronic energy. The Lagrange multiplier enforcing the density normalization constraint is also called the chemical potential, i.e.,

$$\partial \left\{ E[p] - \mu \left(\int p(r)d^3r - N \right) \right\} = 0$$

where N is the number of electrons in the system and μ is the Lagrange multiplier enforcing the constraint. When this variational statement is satisfied, the terms within the curly brackets obey the property

$$\left[\frac{\partial E[p]}{\partial p(\mathbf{r})}\right]_{p=p0} - \mu\left[\frac{\partial N[p]}{\partial p(\mathbf{r})}\right]_{p=p0} = 0$$

where the reference density is the density that minimizes the energy. This expression simplifies to

$$\left[\frac{\partial E[p]}{\partial p(\mathbf{r})}\right]_{p=p0} = \mu$$

The Lagrange multiplier enforcing the constraint is, by construction, a constant; however, the functional derivative is, formally, a function. Therefore, when the density minimizes the electronic energy, the chemical potential has the same value at every point in space. The gradient of the chemical potential is an effective electric field. An electric field describes the force per unit charge as a function of space. Therefore, when the density is the ground state density, the electron density is stationary, because the gradient of the chemical potential (which is invariant with respect to position) is zero everywhere, i.e., all forces are balanced. As the density undergoes a change from a non-ground state density to the ground state density, it is said to undergo a process of chemical potential equalization.

The chemical potential of an atom is sometimes said to be the negative of the atom's electronegativity. Similarly the process of chemical potential equalization is sometimes referred to as the process of *electronegativity equalization*. This connection comes from the Mulliken definition of electronegativity. By inserting the energetic definitions of the ionization potential and electron affinity into the Mulliken electronegativity, it is possible to show that the Mulliken chemical potential is a finite difference approximation of the electronic energy with respect to the number of electrons., i.e.,

$$\mu_{\text{Mulliken}} = -X_{\text{Mulliken}} = -\frac{IP+EA}{2} = \left[\frac{\partial E[N]}{\partial N}\right]_{N=N_0}$$

where *IP* and *EA* are the ionization potential and electron affinity of the atom, respectively.

The Values of the Chemical Potential

For standard conditions (T = 298.15 K; p = 101,325 Pa) the values of the chemical potential are tabulated, see under "Weblinks". If the chemical potential is known in a certain state (e.g. for standard conditions), then it can be calculated in linear approximation for pressures and temperatures in the vicinity of this state:

$$\mu(T) = \mu(T_0) + a(T - T_0)$$

and

$$\mu(p) = \mu(p_0) + \beta(p - p_0)$$

Here

$$\alpha = \left(\frac{\partial \mu}{\partial T}\right)_{p,n}$$

is the temperature coefficient and

$$\beta = \left(\frac{\partial \mu}{\partial p}\right)_{T,n}$$

is the pressure coefficient.

With the Maxwell relations

$$\left(\frac{\partial \mu}{\partial p}\right)_{p,n} = -\left(\frac{\partial S}{\partial n}\right)_{T,p}$$

and

$$\left(\frac{\partial \mu}{\partial p}\right)_{T,n} = -\left(\frac{\partial V}{\partial n}\right)_{T,p}$$

it follows that the temperature coefficient is equal to the negative molar entropy and the pressure coefficient is equal to the molar volume.

Fundamental Particle Chemical Potential

In recent years, thermal physics has applied the definition of chemical potential to systems in particle physics and its associated processes. In general, chemical potential measures the tendency of particles to diffuse. This characterization focuses on the chemical potential as a function of spatial location. Particles tend to diffuse from regions of high chemical potential to those of low chemical potential Being a function of internal energy, chemical potential applies equally to both fermion and boson particles, That is, in theory, any fundamental particle can be assigned a value of chemical potential, depending upon how it changes the internal energy of the system into which it is introduced. The application of chemical potential concepts for systems at absolute zero has significant appeal.

For relativistic systems, *i.e.*, systems in which the rest mass is much smaller than the equivalent thermal energy, the chemical potential is related to symmetries and charges. Each conserved quantity is associated with a chemical potential.

In a gas of photons in equilibrium with massive particles, the number of photons is not conserved, and so in this case, the chemical potential is zero. Similarly, for a gas of phonons, there is also no chemical potential. However, if the temperature of such a system were to rise above the threshold for pair production of electrons, then it might be sensible to add a chemical potential for the electrical charge. This would control the electric charge density of the system, and hence the excess of electrons over positrons, but not the number of photons. In the context in which one meets a phonon gas, temperatures high enough to pair produce other particles are seldom relevant. QCD matter is the prime example of a system in which many such chemical potentials appear.

15

Chemical Equilibrium

In a chemical process, chemical equilibrium is the state in which the chemical activities or concentrations of the reactants and products have no net change over time. Usually, this would be the state that results when the forward chemical process proceeds at the same rate as their reverse reaction. The reaction rates of the forward and reverse reactions are generally not zero but, being equal, there are no net changes in any of the reactant or product concentrations. This process is called dynamic equilibrium.

Introduction

In a chemical reaction, when reactants are mixed together in a reaction vessel (and heated if needed), the whole of reactants do not get converted into the products. After some time (which may be shorter than millionths of a second or longer than the age of the universe), the opposing reactions will have equal reaction rates, creating a dynamic equilibrium in which the ratio between reactants and products will appear fixed. This is called chemical equilibrium.

The concept of chemical equilibrium was developed after Berthollet (1803) found that some chemical reactions are reversible. For any reaction such as

$$\alpha A + \beta B \rightleftharpoons \sigma S + \tau T$$

to be at equilibrium the rates of the forward and backward (reverse) reactions have to be equal. In this chemical equation with harpoon arrows pointing both ways to indicate equilibrium, A and B are reactant chemical species, S and T are product species, and α, ß, σ, and τ are the stoichiometric coefficients of the respective reactants and products. The equilibrium position of a reaction is said to lie far to the right if, at equilibrium, nearly all the reactants are used up and far to the left if hardly any product is formed from the reactants.

Guldberg and Waage (1865), building on Berthollet's ideas, proposed the law of mass action:

forward reaction rate = $k_+ A^\alpha B^\beta$

backward reaction rate = $k_- S^\sigma T^\tau$

where A, B, S and T are active masses and k_+ and k_- are rate constants. Since forward and backward rates are equal:

$$k_+ \{A\}^\alpha \{B\}^\beta = k_- \{S\}^\sigma \{T\}^\tau$$

and the ratio of the rate constants is also a constant, now known as an equilibrium constant.

$$K = \frac{k_+}{k_-} = \frac{\{S\}^\sigma \{T\}^\tau}{\{A\}^\alpha \{B\}^\beta}$$

By convention the products form the numerator. However, the law of mass action is valid only for concerted one-step reactions that proceed through a single transition state and is not valid in general because rate equations do not, in general, follow the stoichiometry of the reaction as Guldberg and Waage had proposed (see, for example, nucleophilic aliphatic substitution by SN1 or reaction of hydrogen and bromine to form hydrogen bromide). Equality of forward and backward reaction rates, however, is a necessary condition for chemical equilibrium, though it is not sufficient to explain why equilibrium occurs.

Despite the failure of this derivation, the equilibrium constant for a reaction is indeed a constant, independent of the activities of the various species involved, though it does depend on temperature as observed by the van 't Hoff equation. Adding a catalyst will affect both the forward reaction and the reverse reaction in the same way and will not have an effect on the equilibrium constant. The catalyst will speed up both reactions thereby increasing the speed at which equilibrium is reached in time reactions do occur at the molecular level. For example, in the case of ethanoic acid dissolved in water and forming ethanoate and hydronium ions.

$$CH_3CO_2H + H_2O \rightleftharpoons CH_3CO^-_2 + H_3\ O^+$$

a proton may hop from one molecule of ethanoic acid on to a water molecule and then on to an ethanoate ion to form another molecule of ethanoic acid and leaving the number of ethanoic acid molecules unchanged. This is an example of dynamic equilibrium. Equilibria, like the rest of thermodynamics, are statistical phenomena, averages of microscopic behavior.

Le Chatelier's principle (1884) is a useful principle that gives a qualitative idea of an equilibrium system's response to changes in reaction conditions. *If a dynamic equilibrium is disturbed by changing the conditions, the position of equilibrium moves to counteract the change.* For example, adding more S from the outside will cause an excess of products, and the system will try to counteract this by increasing the reverse reaction and pushing the equilibrium point backward (though the equilibrium constant will stay the same).

If mineral acid is added to the ethanoic acid mixture, increasing the concentration of hydronium ion, the amount of dissociation must decrease as the reaction is driven to the left in accordance with this principle. This can also be deduced from the equilibrium constant expression for the reaction:

$$K = \frac{\{CH_3CO_2^-\}\{H_3O_+\}}{\{CH_3CO2H\}\{H_2O\}}$$

if $\{H_3O^+\}$ increases $\{CH_3CO_2H\}$ must increase and $\{CH_3CO_2^-\}$ must decrease.

A quantitative version is given by the reaction quotient.

J.W. Gibbs suggested in 1873 that equilibrium is attained when the Gibbs energy of the system is at its minimum value (assuming the reaction is carried out under constant pressure). What this means is that the derivative of the Gibbs energy with respect to reaction coordinate (a measure of the extent of reaction that has occurred, ranging from zero for all reactants to a maximum for all products) vanishes, signalling a stationary point. This derivative is usually called, for certain technical reasons, the Gibbs energy change. This criterion is both necessary and sufficient. If a mixture is not at equilibrium, the liberation of the excess Gibbs energy (or Helmholtz energy at constant volume reactions) is the "driving force" for the composition of the mixture to change until equilibrium is reached. The equilibrium constant can be related to the standard Gibbs energy change for the reaction by the equation

$$\Delta_r G^\Theta = -RT \ln K_{eq}$$

where R is the universal gas constant and T the temperature.

When the reactants are dissolved in a medium of high ionic strength the quotient of activity coefficients may be taken to be constant. In that case the concentration quotient, K_c,

$$K_c = \frac{[S]^{\sigma}[T]^{\tau}}{[A]^{\sigma}[B]^{\beta}}$$

where [A] is the concentration of A, etc., is independent of the analytical concentration of the reactants. For this reason, equilibrium constants for solutions are usually determined in media of high ionic strength. K_c varies with ionic strength, temperature and pressure (or volume). Likewise K_p for gases depends on partial pressure. These constants are easier to measure and encountered in high-school chemistry courses.

Thermodynamics

The relationship between the Gibbs energy and the equilibrium constant can be found by considering chemical potentials. At constant temperature and pressure the fonction **G** Gibbs free energy for the reaction, depends only with the extent of reaction: ε and can only decrease according to the second law of thermodynamics. It means that the derivative of **G** with ε must be negative if the reaction happens; at the equilibrium the derivative being equal to zero.

$$\left(\frac{dG}{d\xi}\right)_{T,p} = 0 \text{ equilibrium}$$

At constant volume, one must consider the Helmholtz free energy for the reaction: **A**.

The constant volume case is important in geochemistry and atmospheric chemistry where pressure variations are significant. Note that, if reactants and products were in standard state (completely pure), then there would be no reversibility and no equilibrium. The mixing of the products and reactants contributes a large entropy (known as entropy of mixing) to states containing equal mixture of products and reactants. The combination of the standard Gibbs energy change and the Gibbs energy of mixing determines the equilibrium state.

In general an equilibrium system is defined by writing an equilibrium equation for the reaction

$$\alpha A + \beta B \qquad \sigma S + \tau T$$

In order to meet the thermodynamic condition for equilibrium, the Gibbs energy must be stationary, meaning that the derivative of G with respect to the extent of reaction: ε, must be zero. It can be shown that in this case, the sum of chemical potentials of the products is equal to the sum of those corresponding to the reactants. Therefore, the sum of the Gibbs energies of the reactants must be the equal to the sum of the Gibbs energies of the products.

$$\alpha\mu_A + \beta\mu_B = \sigma\mu_S + \tau\mu_T$$

where μ is in this case a partial molar Gibbs energy, a chemical potential. The chemical potential of a reagent A is a function of the activity, {A} of that reagent.

$\mu A = \mu_A^{\ominus} + RT \ln\{A\} \mu_A^{\Theta}$ is the standard chemical potential).

Substituting expressions like this into the Gibbs energy equation:

$dG=Vdp - SdT + \sum_{i=l}^{k} \mu_i dN_i$ in the case of a closed system.

Now

$dN_i = v_i d\xi$ (v_i corresponds to the stoechiometric coefficient and dε is the differential of the extent of reaction).

At constant pressure and temperature is obtained:

$$\left(\frac{dG}{d\xi}\right)_{T,P} = \sum_{i=1}^{k} \mu_i \nu_i = \Delta_r G_{T,P}$$

which corresponds to the Gibbs free energy change for the reaction.

This results in:

$$\Delta_r G_{T,P} = \sigma\mu_S + \tau\mu_T - \sigma\mu_A - \beta\mu_B$$

By substituting the chemical potentials:

$$\Delta_r G_{T,P} = \left(\sigma\mu_S^\Theta + \tau\mu_T^\Theta\right) - \left(\sigma\mu_A^\Theta - \beta\mu_B^\Theta\right) + \left(\sigma RT \ln\{S\} + \tau RT \ln\{T\}\right) - \left(\sigma RT \ln\{A\} + \beta RT \ln\{B\}\right)$$

the relationship becomes:

$$\Delta rGTp = \sum_{i=l}^{k} \mu_i^\Theta \nu_i + RT \ln \frac{\{S\}^\sigma \{T\}^\tau}{\{A\}^\alpha \{B\}^\beta}$$

$$\sum_{i=l}^{k} \mu_i^\Theta \nu_i = \Delta_r G^\Theta :$$

which is the standard Gibbs energy change for the reaction. It is a constant at a given temperature, which can be calculated, using thermodynamical tables.

$$RT \ln \frac{\{S\}^\sigma \{T\}^\tau}{\{A\}^\alpha \{B\}^\beta} = RT \ln Q_\tau$$

(Q_r is the reaction quotient when the system is not at equilibrium).

Therefore

$$\left(\frac{dG}{d\xi}\right)_{T,p} = \Delta_r GT_{,p} = \Delta_r G^\Theta + RT \ln Q_r$$

At equilibrium $\left(\frac{dG}{d\xi}\right)_{T,p} = \Delta_r GT_{,p} = 0$

$Q_r = K_{eq}$; the reaction quotient becomes equal to the equilibrium constant.

Leading to:

$$0 = \Delta_r G^{\Theta} + RT \ln K_{eq}$$

and

$$\Delta_r G^{\Theta} = -RT \ln K_{eq}$$

Obtaining the value of the standard Gibbs energy change, allows the calculation of the equilibrium constant.

Addition of Reactants/Products

For a reactionnal system at equilibrium:

$$Q_r = K_{(eq)};\ \xi = \xi_{eq}$$

If are modified activities of constituants, the value of the reaction quotient changes and becomes different from the equilibrium constant: $Q_r \neq K_{(eq)}$

$$\left(\frac{dG}{d\xi}\right)_{T,p} = \Delta_r G^{\Theta} + RTlnQr$$

and

$$\Delta_r G^{\Theta} = -RTlnK_{(eq)}$$

then

$$\left(\frac{dG}{d\xi}\right)_{T,p} = -RTln\left(\frac{Qr}{K_{(eq)}}\right)$$

If activity of a reagent ***i*** increases

$$Qr = \frac{\Pi\left(a_j\right)^{vj}}{\Pi\left(a_i\right)^{vj}}$$

then

$$Qr < K_{(eq)} \text{and} \left(\frac{dG}{d\xi}\right)_{T,p} < 0$$

The reaction will shift to the right (i.e. in the forward direction, and thus more products will form).

If activity of a product ***i*** increases then

$$Qr < K_{(eq)} \text{and} \left(\frac{dG}{d\xi}\right)_{T,p} < 0$$

The reaction will shift to the left (i.e. in the reverse direction, and thus less products will form).

Note that activities and equilibrium constants are dimensionless numbers.

Treatment of Activity

The expression for the equilibrium constant can be rewritten as the product of a concentration quotient, K_c and an activity coefficient quotient, Γ

$$K = \frac{[S]^{\sigma}[T]^{\tau}\ldots}{[A]^{\alpha}[B]^{\beta}\ldots} \times \frac{\Upsilon S^{\sigma}\Upsilon T^{\tau}\ldots}{\Upsilon A^{\alpha}\Upsilon B^{\beta}\ldots} = K_c\Gamma$$

[A] is the concentration of reagent A, etc. It is possible in principle to obtain values of the activity coefficients, γ. For solutions, equations such as the Debye-Hückel equation or extensions such as Davies equation or Pitzer equations may be used. However this is not always possible. It is common practice to assume that Γ is a constant, and to use the concentration quotient in place of the thermodynamic equilibrium constant. It is also general practice to use the term equilibrium constant instead of the more accurate concentration quotient. This practice will be followed here.

For reactions in the gas phase partial pressure is used in place of concentration and fugacity coefficient in place of activity coefficient. In the real world, for example, when making ammonia in industry, fugacity coefficients must be taken into account. Fugacity, *f*, is the product of partial pressure and fugacity coefficient. The chemical potential of a species in the gas phase is given by

$$\mu = \mu^{\Theta} + RT \ln\left(\frac{f}{bar}\right) + RT \ln \gamma$$

so the general expression defining an equilibrium constant is valid for both solution and gas phases.

Justification for the use of Concentration Quotients

In aqueous solution, equilibrium constants are usually determined in the presence of an "inert" electrolyte such as sodium nitrate $NaNO_3$ or Potassium perchlorate $KClO_4$. The ionic strength, *I*, of a solution containing a dissolved salt, X^+Y^-, is given by

$$I = \frac{1}{2}\left(c_x z_x^2 + c_y z_y^2 + \sum_{i=1}^{n} c_i z_i^2\right)$$

where *c* stands for concentration, *z* stands for ionic charge and the sum is taken over all the species in equilibrium. When the concentration of dissolved salt is much higher than the analytical concentrations of the reagents, the ionic strength is effectively constant. Since activity coefficients depend on ionic strength the activity coefficients of the species are effectively independent of concentration. Thus, the assumption that Γ is constant is justified. The concentration quotient is a simple multiple of the equilibrium constant

$$K_c = \frac{K}{T}$$

However, K_c will vary with ionic strength. If it is measured at a series of different ionic strengths the value can be extrapolated to zero ionic strength The concentration quotient obtained in this manner is known, paradoxically, as a thermodynamic equilibrium constant.

To use a published value of an equilibrium constant in conditions of ionic strength different from the conditions used in its determination, the value should be adjusted

Metastable Mixtures

A mixture may be appear to have no tendency to change, though it is not at equilibrium. For example, a mixture of SO_2 and O_2 is metastable as there is a kinetic barrier to formation of the product, SO_3.

$$2SO_2 + O_2 \rightleftharpoons 2SO_3$$

The barrier can be overcome when a catalyst is also present in the mixture as in the contact process, but the catalyst does not affect the equilibrium concentrations.

Likewise, the formation of bicarbonate from carbon dioxide and water is very slow under normal conditions.

$$CO_2 + 2H_2O \rightleftharpoons HCO_3 + H_3O^+$$

but almost instantaneous in the presence of the catalytic enzyme carbonic anhydrase.

Pure Compounds in Equilibria

When pure substances (liquids or solids) are involved in equilibria they do not appear in the equilibrium equation.

Applying the general formula for an equilibrium constant to the specific case of ethanoic acid one obtains

$$CH_3CO_2H + H_2O \rightleftharpoons CH_3CO_2 + H_3O^+$$

$$K_c = \frac{\left[CH_3CO_2^-\right]\left[H_3O^+\right]}{\left[CH_3CO_2H\right]\left[H_2O\right]}$$

It may be assumed that the concentration of water is constant. This assumption will be valid for all but very concentrated solutions. The equilibrium constant expression is therefore usually written as

$$K - \frac{\left[CH_3CO_2^-\right]\left[H_3O^+\right]}{\left[CH_3CO_2H\right]}$$

where now

$K = Kc * [H_2O]$

a constant factor is incorporated into the equilibrium constant.

A particular case is the self-ionization of water itself

$H_2O + H_2O = H_3O^+ + OH^-$

The self-ionization constant of water is defined as

$K_w = [H^+]\ [OH^-]$

It is perfectly legitimate to write [H^+] for the hydronium ion concentration, since the state of solvation of the proton is constant (in dilute solutions) and so does not affect the equilibrium concentrations. K_w varies with variation in ionic strength and/or temperature.

The concentrations of H^+ and OH^- are not independent quantities. Most commonly [OH^-] is replaced by $Kw[H^+]^{-1}$ in equilibrium constant expressions which would otherwise hydroxide.

Solids also do not appear in the equilibrium equation. An example is the Boudouard reaction

$$2CO \rightleftharpoons CO_2 + C$$

for which the equation (without solid carbon) is written as:

$$K_c = \frac{[CO_2]}{[CO]^2}$$

Multiple Equilibria

Consider the case of a dibasic acid H_2A. When dissolved in water, the mixture will contain H_2A, HA^- and A^{2-}. This equilibrium can be split into two steps in each of which one proton is liberated.

$$H_2A = HA^- + : K_1 = \frac{[HA^-][H^+]}{[H_2A]}$$

$$HA^- = A^{2-} + H+ : k_2 = \frac{[H2^-][H^+]}{[HA^-]}$$

K_1 and K_2 are examples of *stepwise* equilibrium constants. The *overall* equilibrium constant,ß*D*, is product of the stepwise constants.

$$H_2A = A^{2-} + 2H^+ : \beta b = \frac{[A^{2-}][H^+]^2}{[H_2A]} = K_1K_2$$

Note that these constants are dissociation constants because the products on the right hand side of the equilibrium expression are dissociation products. In many systems, it is preferable to use association constants.

$$A^{2-} + H^+ = HA^- : \beta_1 = \frac{[HA^-]}{[A^{2-}][H^+]}$$

$$A^{2-} + 2H^+ = H_2A : \beta_2 = \frac{[H_2A]}{[A^{2-}][H^+]^2}$$

$ß_1$ and $ß_2$ are examples of association constants. Clearly $ß_1 = 1/K_2$ and $ß_2 = 1/ß_D$; lg $ß_1$ = pK_2 and lg $ß_2$ = $pK_2 + pK_1$

Effect of Temperature Change on an Equilibrium Constant

The effect of changing temperature on an equilibrium constant is given by the van 't Hoff equation

$$\frac{d \ln K}{dT} = \frac{\Delta H_m^\Theta}{RT^2}$$

Thus, for exothermic reactions, (?H is negative) *K* decreases with an increase in temperature, but, for endothermic reactions, (?H is positive) *K* increases with an increase temperature. An alternative formulation is

$$\frac{d \ln K}{d(1/T)} = -\frac{\Delta H_m^\Theta}{R}$$

At first sight this appears to offer a means of obtaining the standard molar enthalpy of the reaction by studying the variation of K with temperature. In practice, however, the method is unreliable because error propagation almost always gives very large errors on the values calculated in this way.

Types of Equilibrium and some Applications

- In the gas phase. Rocket engines.
- The industrial synthesis such as ammonia in the Haber-Bosch process (depicted right) takes place through a succession of equilibrium steps including adsorbtion processes.
- Atmospheric chemistry
- Seawater and other natural waters: Chemical oceanography distribution between two phases:
- LogD-Distribution coefficient: Important for pharmaceuticals where lipophilicity is a significant property of a drug.

1. Liquid-liquid extraction, Ion exchange, Chromatography.
2. Solubility product.
3. Uptake and release of oxygen by haemoglobin in blood.
4. *Acid/base equilibria:* Acid dissociation constant, hydrolysis buffer solutions, indicators, acid-base homeostasis.

- *Metal-ligand complexation:* sequestering agents, chelation therapy, MRI contrast reagents, Schlenk equilibrium.
- *Adduct formation:* Host-guest chemistry, supramolecular chemistry, molecular recognition, dinitrogen tetroxide.
- In *certain oscillating reactions*, the approach to equilibrium is not asymptotically but in the form of a damped oscillation.
- *The related Nernst equation* in electrochemistry gives the difference in electrode potential as a function of redox concentrations.

- When molecules on each side of the equilibrium are able to further react irreversibly in secondary reactions, the final product ratio is determined according to the Curtin-Hammett principle.

In these applications, terms such as stability constant, formation constant, binding constant, affinity constant, association/dissociation constant are used. In biochemistry, it is common to give units for binding constants, which serve to define the concentration units used when the constant's value was determined.

Composition of an Equilibrium Mixture

When the only equilibrium is that of the formation of a 1:1 adduct as he composition of a mixture, there are any number of ways tat the composition of a mixture can be calculated.

There are three approaches to the general calculation of the composition of a mixture at equilibrium.

- The most basic approach is to manipulate the various equilibrium constants until the desired concentrations are expressed in terms of measured equilibrium constants (equivalent to measuring chemical potentials) and initial conditions.
- Minimize the Gibbs energy of the system.
- Satisfy the equation of mass balance. The equations of mass balance are simply statements that demonstrate that the total concentration of each reactant must be constant by the law of conservation of mass.

Solving the Equations of Mass-balance

In general, the calculations are rather complicated. For instance, in the case of a dibasic acid, H_2A dissolved in water the two reactants can be specified as the conjugate base, A^{2-}, and the proton, H^+. The following equations of mass-balance could apply equally well to a base such as 1,2-diaminoethane, in which case the base itself is designated as the reactant A:

$$T_A = [A] + [HA] + [H_2A]$$

$$T_H = [H] + [HA] + 2[H_2A] - [OH]$$

With T_A the total concentration of species A. Note that it is customary to omit the ionic charges when writing and using these equations.

When the equilibrium constants are known and the total concentrations are specified there are two equations in two unknown "free concentrations" [A] and [H]. This follows from the fact that $[HA] = ß_1[A][H]$, $[H_2A] = ß_2[A][H]^2$ and $[OH] = K_w[H]^{-1}$

$$T_A = [A] + \beta_1 [A] [H] + \beta_2 [A] [H]^2$$

$$T_H = [H] + \beta_1 [A] [H] + 2\beta_2 [A] [H]^2 - K_w [H]^{-1}$$

so the concentrations of the "complexes" are calculated from the free concentrations and the equilibrium constants. General expressions applicable to all systems with two reagents, A and B would be

$$T_A = [A] + \sum_i pi\beta i[A]^{pi}[B]^{qi}$$

$$T_B = [B] + \sum_i qi\beta i[A]^{pi}[B]^{qi}$$

It is easy to see how this can be extended to three or more reagents.

Composition for polybasic acids as a function of pH

The composition of solutions containing reactants A and H is easy to calculate as a function of p[H]. When [H] is known, the free concentration [A] is calculated from the mass-balance equation in A. Here is an example of the results that can be obtained.

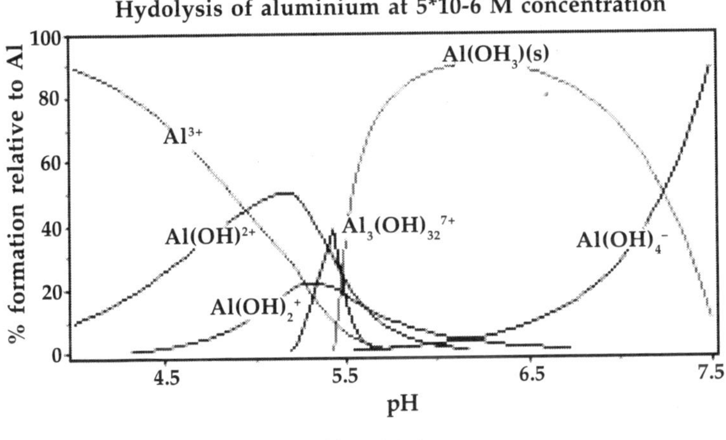

Fig. 15.1

This diagram (15.1) for the hydrolysis of the aluminium Lewis acid Al3+aq shows the species concentrations for a 5×10^{-6}M solution of an aluminium salt as a function of pH. Each concentration is shown as a percentage of the total aluminium.

Solution Equilibria with Precipitation

The diagram above illustrates the point that a precipitate that is not one of the main species in the solution equilibrium may be formed. At pH just below 5.5 the main species present in a 5μM solution of Al^{3+} are aluminium hydroxides $Al(OH)^{2+}$, $Al(OH)_2^+$ and $Al_{13}(OH)_{32}^{7+}$, but on raising the pH $Al(OH)_3$ precipitates from the solution. This occurs because $Al(OH)_3$ has a very large lattice energy. As the pH rises more and more $Al(OH)_3$ comes out of solution. This is an example of Le Chatelier's principle in action: Increasing the concentration of the hydroxide ion causes more aluminium hydroxide to precipitate, which removes hydroxide from the solution. When the hydroxide concentration becomes sufficiently high the soluble aluminate, $Al(OH)_4^-$, is formed.

Another common instance where precipitation occurs is when a metal cation interacts with an anionic ligand to form an electrically-neutral complex. If the complex is hydrophopbic, it will precipitate out of water. This occurs with the nickel ion Ni^{2+} and dimethylglyoxime, ($dmgH_2$): in this case the lattice energy of the solid is not particularly large, but it greatly exceeds the energy of solvation of the molecule $Ni(dmgH)_2$.

Minimization of Gibbs Energy

At equilibrium, G is at a minimum:

$$dG = \sum_{j=1}^{m} \mu_j dN_j = 0$$

For a closed system, no particles may enter or leave, although they may combine in various ways. The total number of atoms of each element will remain constant. This means that the minimization above must be subjected to the constraints:

$$\sum_{j=1}^{m} a_{ij} N_j = b_i^0$$

where a_{ij} is the number of atoms of element i in molecule j and b^o_i is the total number of atoms of element i, which is a constant, since the system is closed. If there are a total of k types of atoms in the system, then there will be k such equations.

This is a standard problem in optimisation, known as constrained minimisation. The most common method of solving it is using the method of Lagrange multipliers, also known as undetermined multipliers (though other methods may be used).

Define:

$$\mathcal{G} = G + \sum_{i=l}^{k} \lambda_i \left(\sum_{j=1}^{m} a_{ij} N_j - b_i^0 \right) = 0$$

where the λ_i are the Lagrange multipliers, one for each element. This allows each of the N_j to be treated independently, and it can be shown using the tools of multivariate calculus that the equilibrium condition is given by

$$\frac{\partial}{\partial N_j} = 0 \text{ and } \frac{\partial}{\partial \lambda_i} = 0$$

This is a set of $(m+k)$ equations in $(m+k)$ unknowns (the N_j and the λ_i and may, therefore, be solved for the equilibrium concentrations N_j as long as the chemical potentials are known as functions of the concentrations at the given temperature and pressure.

This method of calculating equilibrium chemical concentrations is useful for systems with a large number of different molecules. The use of k atomic element conservation equations for the mass constraint is straightforward, and replaces the use of the stoichiometric coefficient equations.

16

Chemical Reactor

In chemical engineering, chemical reactors are vessels designed to contain chemical reactions. The design of a chemical reactor deals with multiple aspects of chemical engineering. Chemical engineers design reactors to maximize net present value for the given reaction. Designers ensure that the reaction proceeds with the highest efficiency towards the desired output product, producing the highest yield of product while requiring the least amount of money to purchase and operate. Normal operating expenses include energy input, energy removal, raw material costs, labor, etc. Energy changes can come in the form of heating or cooling, pumping to increase pressure, frictional pressure loss (such as pressure drop across a 90o elbow or an orifice plate), agitation, etc.

Overview

There are two main basic vessel types:

- a tank
- a pipe

Both types can be used as continuous reactors or batch reactors. Most commonly, reactors are run at steady-state, but can also be operated in a transient state. When a reactor is first brought back into operation (after maintenance or inoperation) it would be considered to be in a transient state, where key process variables change with time. Both types of reactors may also accommodate one or more solids (reagents, catalyst, or inert materials), but the reagents and products are typically liquids and gases.

There are three main basic models used to estimate the most important process variables of different chemical reactors:

- *batch reactor* model (batch);
- *continuous stirred-tank reactor* model (CSTR); and
- *plug flow reactor* model (PFR).

Furthermore, catalytic reactors require separate treatment, whether they are batch, CST, or PF reactors, as the many assumptions of the simpler models are not valid.

Key process variables include:

- residence time (t, lower case Greek tau)
- volume (V)
- temperature (T)
- pressure (P)
- concentrations of chemical species (C_1, C_2, C_3, . . . Cn)
- heat transfer coefficients (h, U)

Types

CSTR (Continuous Stirred-Tank Reactor)

In a CSTR, one or more fluid reagents are introduced into a tank reactor equipped with an impeller while the reactor effluent is removed. The impeller stirs the reagents to ensure proper mixing. Simply dividing the volume of the tank by the average volumetric flow rate through the tank gives the *residence time*, or the average amount of time a discrete quantity of reagent spends inside the tank. Using chemical kinetics, the reaction's expected percent completion can be calculated. Some important aspects of the CSTR:

- At steady-state, the flow rate in must equal the mass flow rate out, otherwise the tank will overflow or go empty (transient state). While the reactor is in a transient state the model equation must be derived from the differential mass and energy balances.
- The reaction proceeds at the reaction rate associated with the final (output) concentration.
- Often, it is economically beneficial to operate several CSTRs in series. This allows, for example, the first CSTR to operate at a higher reagent concentration and therefore a higher reaction rate. In these cases, the sizes of the reactors may be varied in order to minimize the total capital investment required to implement the process.
- It can be seen that an infinite number of infinitely small CSTRs operating in series would be equivalent to a PFR.

The behaviour of a CSTR is often approximated or modeled by that of a Continuous Ideally Stirred-Tank Reactor (CISTR). All calculations performed with CISTRs assume perfect mixing. If the residence time is 5-10 times the mixing time, this approximation is valid for engineering purposes. The CISTR model is often used to simplify engineering calculations and can be used to describe research reactors. In practice it can only be approached, in particular in industrial size reactors.

PFR (Plug Flow Reactor)

In a PFR, one or more fluid reagents are pumped through a pipe or tube. The chemical reaction proceeds as the reagents travel through the PFR. In this type of reactor, the changing reaction rate creates a gradient with respect to distance traversed; at the inlet to the PFR the rate is very high, but as the concentrations of the reagents decrease and the concentration of the product(s) increases the reaction rate slows. Some important aspects of the PFR:

- All calculations performed with PFRs assume no upstream or downstream mixing, as implied by the term "plug flow".
- Reagents may be introduced into the PFR at locations in the reactor other than the inlet. In this way, a higher efficiency may be obtained, or the size and cost of the PFR may be reduced.
- A PFR typically has a higher efficiency than a CSTR of the same volume. That is, given the same space-time, a reaction will proceed to a higher percentage completion in a PFR than in a CSTR.

For most chemical reactions, it is impossible for the reaction to proceed to 100% completion. The rate of reaction decreases as the percent completion increases until the point where the system reaches dynamic equilibrium (no net reaction, or change in chemical species occurs). The equilibrium point for most systems is less than 100% complete. For this reason a separation process, such as distillation, often follows a chemical reactor in order to separate any remaining reagents or byproducts from the desired product. These reagents may sometimes be reused at the beginning of the process, such as in the Haber process.

Continuous oscillatory baffled reactor (COBR) is a tubular plug flow reactor. The mixing in COBR is achieved by the combination of fluid oscillation and orifice baffles, allowing plug flow to be achieved under laminar flow conditions with the net flow Reynolds number just about 100.

Semi-batch Reactor

A semi-batch reactor is operated with both continuous and batch inputs and outputs. A fermenter, for example, is loaded with a batch, which constantly produces carbon dioxide, which has to be removed continuously. Analogously, driving a reaction of gas with a liquid is usually difficult, since the gas bubbles off. Therefore, a continuous feed of gas is injected into the batch of a liquid. An example of such a reaction is chlorination.

Catalytic Reactor

Although catalytic reactors are often implemented as plug flow reactors, their analysis requires more complicated treatment. The rate of a catalytic reaction is proportional to the amount of catalyst the reagents contact. With a solid phase catalyst and fluid phase reagents, this is proportional to the exposed area, efficiency of diffusion of reagents in and products out, and turbulent mixing or lack thereof. Perfect mixing cannot be assumed. Furthermore, a catalytic reaction pathway is often multi-step with intermediates that are chemically bound to the catalyst; and as the chemical binding to the catalyst is also a chemical reaction, it may affect the kinetics.

The behavior of the catalyst is also a consideration. Particularly in high-temperature petrochemical processes, catalysts are deactivated by sintering, coking, and similar processes.

A common example of a catalytic reactor is the catalytic converter following a motor.

Fluidized Bed Reactor

A fluidized bed reactor (FBR) is a type of reactor device that can be used to carry out a variety of multiphase chemical reactions. In this type of reactor, a fluid (gas or liquid) is passed through a granular solid material (usually a catalyst possibly shaped as tiny spheres) at high enough velocities to suspend the solid and cause it to behave as though it were a fluid. This process, known as fluidization, imparts many important advantages to the FBR. As a result, the fluidized bed reactor is now used in many industrial applications.

Basic Principles

The solid substrate (the catalytic material upon which chemical species react) material in the fluidized bed reactor is typically supported by a porous plate, known as a distributor The fluid is then forced through the distributor up through the solid material. At lower fluid velocities, the solids remain in place as the fluid passes through the voids in the material. This is known as a packed bed reactor. As the fluid velocity is increased, the reactor will reach a stage where the force of the fluid on the solids is enough to balance the weight of the solid material. This stage is known as incipient fluidization and occurs at this minimum fluidization velocity. Once this minimum velocity is surpassed, the contents of the reactor bed begin to expand and swirl around much like an agitated tank or boiling pot of water. The reactor is now a fluidized bed. Depending on the operating conditions and properties of solid phase various flow regimes can be observed in this reactor.

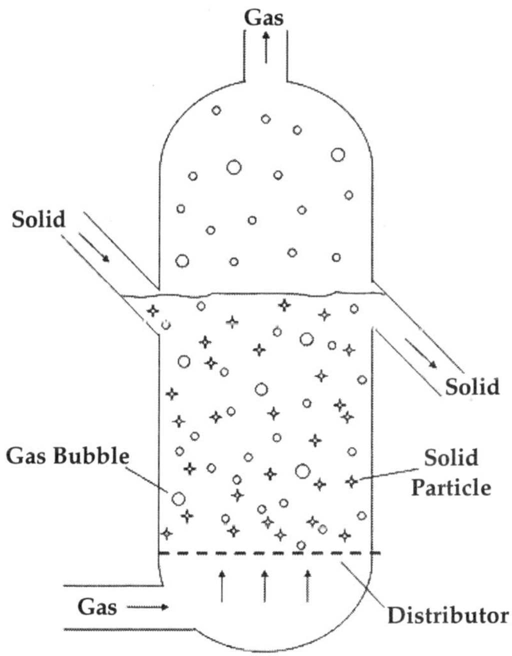

Fig. 16.1

Diagram of Fluidized bad reactor

History and Current Uses

Fluidized bed reactors are a relatively new tool in the chemical engineering field. The first fluidized bed gas generator was developed by Fritz Winkler in Germany in the 1920s One of the first United States fluidized bed reactors used in the petroleum industry was the Catalytic Cracking Unit, created in Baton Rouge, LA in 1942 by the Standard Oil Company of New Jersey (now ExxonMobil) This FBR and the many to follow were developed for the oil and petrochemical industries. Here catalysts were used to reduce petroleum to simpler compounds through a process known as cracking. The invention of this technology made it possible to significantly increase the production of various fuels in the United States.

Today fluidized bed reactors are still used to produce gasoline and other fuels, along with many other chemicals. Many industrially produced polymers are made using FBR technology, such as rubber, vinyl chloride, polyethylene, and styrenes. Various utilities also use FBR's for coal gasification, nuclear power plants, and water and waste treatment settings. Used in these applications, fluidized bed reactors allow for a cleaner, more efficient process than previous standard reactor technologies

Advantages

The increase in fluidized bed reactor use in today's industrial world is largely due to the inherent advantages of the technology

- *Uniform Particle Mixing:* Due to the intrinsic fluid-like behavior of the solid material, fluidized beds do not experience poor mixing as in packed beds. This complete mixing allows for a uniform product that can often be hard to achieve in other reactor designs. The elimination of radial and axial concentration gradients also allows for better fluid-solid contact, which is essential for reaction efficiency and quality.

- *Uniform Temperature Gradients:* Many chemical reactions require the addition or removal of heat. Local hot or cold spots within the reaction bed, often a problem in packed beds, are avoided in a fluidized situation such as an FBR. In other reactor types, these local temperature differences, especially hotspots, can result in product degradation. Thus FBRs are well suited to exothermic reactions. Researchers have also learned that the bed-to-surface heat transfer coefficients for FBRs are high.
- *Ability to Operate Reactor in Continuous State:* The fluidized bed nature of these reactors allows for the ability to continuously withdraw product and introduce new reactants into the reaction vessel. Operating at a continuous process state allows manufacturers to produce their various products more efficiently due to the removal of startup conditions in batch processes.

Disadvantages

As in any design, the fluidized bed reactor does have it draw-backs, which any reactor designer must take into consideration.

- *Increased Reactor Vessel Size:* Because of the expansion of the bed materials in the reactor, a larger vessel is often required than that for a packed bed reactor. This larger vessel means that more must be spent on initial capital costs.
- *Pumping Requirements and Pressure Drop:* The requirement for the fluid to suspend the solid material necessitates that a higher fluid velocity is attained in the reactor. In order to achieve this, more pumping power and thus higher energy costs are needed. In addition, the pressure drop associated with deep beds also requires additional pumping power.

- *Particle Entrainment:* The high gas velocities present in this style of reactor often result in fine particles becoming entrained in the fluid. These captured particles are then carried out of the reactor with the fluid, where they must be separated. This can be a very difficult and expensive problem to address depending on the design and function of the reactor. This may often continue to be a problem even with other entrainment reducing technologies.
- *Lack of Current Understanding:* Current understanding of the actual behavior of the materials in a fluidized bed is rather limited. It is very difficult to predict and calculate the complex mass and heat flows within the bed. Due to this lack of understanding, a pilot plant for new processes is required. Even with pilot plants, the scale-up can be very difficult and may not reflect what was experienced in the pilot trial.
- *Erosion of Internal Components:* The fluid-like behavior of the fine solid particles within the bed eventually results in the wear of the reactor vessel. This can require expensive maintenance and upkeep for the reaction vessel and pipes.

Current Research and Trends

Due to the advantages of fluidized bed reactors, a large amount of research is devoted to this technology. Most current research aims to quantify and explain the behavior of the phase interactions in the bed. Specific research topics include particle size distributions, various transfer coefficients, phase interactions, velocity and pressure effects, and computer modeling. The aim of this research is to produce more accurate models of the inner movements and phenomena of the bed. This will enable scientists and engineers to design better, more efficient reactors that may effectively deal with the current disadvantages of the technology and expand the range of FBR use.

Membrane Reactor

A membrane reactor is a piece of chemical equipment that combines a catalyst-filled reaction chamber with a membrane to add reactants or remove products of the reaction.

Chemical reactors making use of membranes are usually referred to as membrane reactors. The membrane can be used for different tasks:

- Separation
  - Selective extraction of reactants
  - Retention of the catalyst
- Distribution/dosing of a reactant
- Catalyst support (often combined with distribution of reactants)

Membrane reactors are an example for the combination of two unit operations in one step e.g. membrane filtration with the chemical reaction.

Examples

Biological Systems

In biological systems membranes fulfil a number of essential functions. The compartmentalisation of biological cells is achieved by membranes. The semi-permeability allows to separate reactions and reaction environments. A number of enzymes are membrane bound and often mass transport through the membrane is active rather than passive as in artificial membranes allowing the cell to keep up gradients for example by using active transport of protons or water.

The use of a natural membrane is the first example of the utilisation for a chemical reaction. By using the selective permeability of a pigs bladder water could be removed from a condensation reaction to shift the equilibrium position of the reaction towards the condensation products according to the principle of Le Châtelier.

Size Exclusion: Enzyme Membrane Reactor

As enzymes are macromolecules and often differ greatly in size from reactants they can be separated by size exclusion membrane filtration with ultra- or nanofiltration. This is used on industrial scale for the production of enantiopure amino acids by kinetic racemic resolution of chemically derived racemic amino acids. The most prominent example is the production of L-methionine on a scale of 400t/a. The advantage of this method over other forms of immobilisation of the catalyst is that the enzymes are not altered in activity or selectivity as it remains solubilised.

The principle can be applied to all macromolecular catalysts which can be separated from the other reactants by means of filtration. So far, only enzymes have been used to a significant extent.

Reaction Combined with Pervaporation

In P. dense membranes are used for separation. For dense membranes are the separation is governed by the difference of the chemical potential of the reactants in the membrane. The selectivity of the transport through the membrane is given by the different solubility of the materials in the membrane and their diffusivity in the membrane. For example for the selective removal of water by using lipophilic membranes. This can be used to overcome thermodynamic limits of condensation e.g. esterification reactions by removing water.

Dosing: Partial Oxidation of Methane to Methanol

In the star process for the catalytic conversion of from methane from natural gas and air oxygen to methanol by the partial oxidation

$$2CH_4 + O_2 \rightarrow 2CH_3OH.$$

The partial pressure of oxygen has to low to prevent the formation of explosive mixtures and to suppress the successive reaction to carbon monoxide, carbon dioxide and water. This is achieved by using a tubular reactor with an oxygen-selective membrane. The membrane allows the uniform distribution of oxygen as the driving force for the permeation of oxygen through the membrane is the difference in partial pressures on the air side and the methane side.

Selective Removal: Hydrogen

A number of metal membranes are highly hydrogen selective at higher temperatures. Especially palladium and platinium can therefore be used for the production of highly purified hydrogen from steam reforming of gases. The equilibrium limited reaction gives:

$$CH_4 + H_2O \leftrightarrow 3H_2 + CO$$

$$CO + H_2O \leftrightarrow H_2 + CO_2 \text{ or}$$

$$CH_3OH + H_2O \leftrightarrow CO_2 + 3\,H_2$$

Ultra pure hydrogen, generated from these reactions, is extracted by use of thin dense metallic membranes that are 100% selective to hydrogen. The mechanism of the transport is the separation of hydrogen into protons and electrons at the surface and recombination on the filtrate or raffinate side. Other high temperature membranes are being considered for hydrogen generation where the purity requirements are not as great; for example for clean coal power generation. Hydrogen, produced from coal gas in the membrane reactor would be used for power generation, while the carbon dioxide would remain at high pressure for carbon capture and storage.

An alternative application of membrane reactors, developed at University Laval was to convert methane into benzene by the following reaction:

$$6\,CH_4 \rightarrow \text{benzene} + 9\,H_2$$

As with the other reactions, hydrogen extraction drives the conversion forward, but for this reaction, the desired product is the benzene, and not the hydrogen.

Benefits

Making a gaseous product in a membrane reactor generally affects the way that pressure affects the extent of reaction at thermodynamic pseudo-euilibrium. In an ordinary flow reactor, the composition of the exhaust gas is determined by the composition of the feed gas and the extent of reaction. As a result, at pseudo-equilibrium, the extent of reaction is entirely determined by the feed composition and the exhaust equilibrium constant, the latter being determined by the temperature and pressure of the exhaust. In a membrane reactor, the partial pressure of the components at psueudo-equilibrium are not uniquely determined by the total pressure, exit temperature, and feed composition. There is also a significant (and beneficial) effect that derives from the controlled removal of a product or addition of reactant.

Index